ACHIEVING YOUR POTENTIAL

A GUIDE FOR SOFTWARE ENGINEERS

ERROL HASSALL

Tellwell Talent
www.tellwell.ca

ISBN
978-0-2288-8857-4 (Hardcover)
978-0-2288-8856-7 (Paperback)
978-0-2288-8858-1 (eBook)

TABLE OF CONTENTS

Dedication v
Introduction vii

SECTION 1
MY EXPERIENCE...1

My First Year 3
The Good...8
The Challenges of My First Year 15
The Ugly...21
What I Expected vs What Happened 26
You Can Only Control Yourself.................................... 33

SECTION 2
CAREER...35

Interviewing...37
Career Path..48
Taking the Advice of More Experienced Developers.......... 63
Taking the Advice of Less Experienced Developers........... 68
Work-Life Balance... 69
Working with Great People and in Great Teams.................74

SECTION 3
MASTER YOUR CRAFT 81

Goals..83
Ask Questions 89
Take Responsibility... 92
Be Careful What Company You Choose 100
Life Cycle of a Project 116
Professional Software Engineer.................................122
Clean Code..127

Code reviews 132

Testing.. 138

Continuous Integration and Continuous
    Deployment (CI/CD).. 144

Planning Things Before You Start 147

How to Learn Effectively.. 151

Always Learn New Things 156

Finish What You Start... 159

Don't Rush 163

Be Humble 165

Teamwork...170

Be a Rubber Ducky, Find a Rubber Ducky 175

Write Code for Others, Not Yourself 180

Team Dependencies 183

Marketing Yourself... 186

SECTION 4

THE REST 191

Take Breaks 193

Get a Mentor.. 196

Be a Mentor.. 200

Take Care of Your Health 204

Conclusion 215

Book Recommendations...223

DEDICATION

To my wife Casey, the strongest person I know. She gave me the courage to write and, most importantly, publish this book. Without her, my life would have no meaning.

INTRODUCTION

I left high school with no clue about where I was going next. My teenage years had been devoted to becoming an NBA basketball player, but when I hit 17, I knew that goal would never manifest; I was lost. I spent years on this goal, only to realise I simply wasn't good enough and time had run out. Furthermore, I didn't have a backup plan. What was I to do? I enjoyed mathematics immensely, achieving reasonable results in high school, even in the toughest classes, so I thought I might go into the mathematics field.

On the other hand, I loved computers, I had built multiple, and you could always find me tinkering with them. I was constantly looking at the latest parts and dreaming about owning them, so I thought perhaps I could become a programmer. In my second to last year in high school, I tried programming a handful of times, dabbling in bits and pieces at home but not getting anywhere in particular. For some reason, I thought I could build a social network for video games, from scratch, without even knowing HTML. That was a project that didn't last long.

Fast forward to my final year. I enrolled in a university course, which happened to be a maths class, and if I completed it, I would be awarded extra credit for my final year and extra credits for university. I thought to myself, *well, I'm already doing two maths classes, so let's make it three.* I applied for it and got rejected as my grades, especially in maths, weren't spectacular, which was the admission requirement into a first-year university mathematics class. As a consolation prize, I was placed in a programming class. Still, to this day, I can't work out why, but I was. Honestly, it was the second-biggest stroke of luck after meeting my wife. The course was a 12-week structured introduction to algorithms with no prior programming knowledge. Let me stress that last part: no prior programming knowledge. I had no programming knowledge, so I thought this was great. We had class from 6 pm until 8 pm each Wednesday night at Monash University. My dad dropped me off in the early days of March, so it was still light outside at that time of night in Australia. I walked to the room, which required quite a bit of effort as the campus was huge, and waited outside the room. As the small hallway I stood in slowly filled up with other 17-year-olds, who were too nervous to talk to each other, I was reassured that I was actually in the right place. The lecturer, or should I really say the tutor, rocked up and let us in. The room was a sea of computers arranged in groups of four. Two computers facing another two, he asked us to pick any computer, and I did, sitting by myself at first. The next two hours were the most difficult of my life; I had absolutely no clue what was happening. It hit 8 pm, and it was all over. I packed up my things and made my way to the door. I walked to the car where my dad was picking me up and got in.

'How was it, son?' said dad.

'Yeah, great,' I lied.

It was terrifying, my first real computer science subject, and it was intense. The next week, we got our hands dirty with some programming in a tool called 'Scratch.' This was an interactive block-based programming language designed for children. It was still much too advanced for me; I didn't even know what a variable or a loop was or how I would string together multiple concepts. These details were glossed over quickly, and to my horror, I had no idea what was happening. I just sat there screaming internally, freaking out and generally having a bad time. As I was internally screaming, I glanced over at the person next to me; his screen was already filled with blocks, much to my horror. My internal screaming only intensified. All these people were just about done, and I couldn't even place the first block. How stupid was I? I tried to use this person's screen, cheating already or, as I liked to call it, 'guidance.' I kept peaking over at what he had done and would place another block with absolutely no idea what it was doing each time. After 15 minutes or so, the tutor stopped the class, assuming everyone had completed the task. I, of course, did not, but the person next to me certainly had. I plucked up the courage to ask him a few questions, and we quickly hit it off. We talked a lot. James was doing the same subjects in high school, and he had built quite a few websites, even an app. As James showed me the sites he had built, I sat there in amazement at the god-like talent this boy of 17 possessed. I was shocked that someone my age had done anything programming-related. He even took a programming subject in high school, whereas my school didn't even offer one. I was awestruck, to say the least, which became a common theme for me.

The subject was broken up into 12 weeks, with one major assignment broken up into three parts and one exam. The assignment was to build a movie database/social network. Essentially, we were to create an algorithm that could recommend new movies and friends based on the movies the users had already indicated to enjoy. I didn't know what a variable was, let alone have the skill to build something of this magnitude. I hardly understood the assignment, if I'm being completely honest, but more on this later. It got to about week three when it was clear that I had no idea what was going on. However, I found out that others also didn't because the tutor created a special intensive session. The tutor would run you through more of the basics of the subject, mostly about how to program, and provide some resources. It was a way of getting people without knowledge up to speed. It was a live Skype call with about six people, excluding myself and the tutor. The lesson was one hour every Thursday. It was horrifying; he would write up some pseudocode and then ask people questions about it, completely on the spot and more to the point, in front of everyone else. He asked me a few questions, and every time I said the same thing, 'I'm not sure.' It was terrifying and embarrassing, and I wanted nothing more than to leave. I did the most rational thing and never attended another one. The tutor approached me and said it would be very beneficial, but I told him I understood it and would be alright. But I was not alright; I was utterly useless. I chose to pretend I knew the concepts over going to those weekly sessions.

When week eight rocked up, the final assignment was due. Of course, I hadn't done anything because I had no idea what to do. We had been given this program, the same one from before, where we could drag and drop blocks. It also had a graph database function, where we could create a node, such as a person, and then attach it to the movie *Blade Runner.* This assignment was supposed to be a well-documented approach to creating an algorithm that could match users based on their movie preferences. Sixty per cent was designing and documenting, and 40% was code. I spent four hours creating nodes of users and the movies they liked manually, dragging, dropping, giving them a name and some movies, and then wiring them up to another user. I did this for about one hundred users because I couldn't code at all. I ended up getting 60% on this assignment, which meant that if you removed the coding, I would get 100%. I'm still convinced that the tutor just felt sorry for me. The final exam approached, and I had studied my ass off with my newfound resolve; I had gotten 60% on an assignment I expected to fail. James had somewhat studied, but for the most part, he was confident. We met at the exam and smashed it out. James thought he did reasonably well, and I thought I had done surprisingly well. However, I failed the subject, scoring 45% overall. That's the story of my first computer science class. I failed miserably, and my friend got 94%, top of the class. We are still great friends to this day.

My final year of high school was rapidly coming to an end, yet I still hadn't picked what I wanted to do in university. My girlfriend suggested computer science, as I loved computers, and she warned me not to go into mathematics as I wouldn't enjoy it. More importantly, she suggested that I get a job. I decided not to listen to her and went and did mathematics. After my first semester, I hated everything about it. They crammed one year in high school into 12 weeks in university, and I didn't learn anything or enjoy it. I swapped to computer science, and my girlfriend was right, as she usually is, which is probably why she's now my wife.

I can sum up my Computer Science degree quickly: pass and credits. Almost all my grades fell between these two marks. I ended up with roughly a credit average, nothing special, and I only had two subjects in which I got a high distinction. One was my final year project, a group project that didn't fare so well, with two people not doing anything. Out of the fallout, I got an HD (High Distinction) based on my contribution to the project. The only other HD was from a law subject, which was dead easy because all you had to do was follow a formula. It was an open-book exam, and my girlfriend had already done it and given me all her notes. I also failed a subject and almost another, so let's be honest, I had a pretty standard degree, maybe even below standard. Before I finished, I applied to roughly 50 jobs, only getting one response. That one response was from a recruiter who was clearly not good at his job, as he thought I was a senior developer. I couldn't tell you what I applied for because I just applied for everything that had Dot Net. At the time, Dot Net was the only language I knew. However, I made a great friend who, after finishing my degree, put me in touch with someone looking to hire. The contact got me the interview, and I got the job. The ecstatic feeling of securing your first job in your new career is something I won't forget anytime soon. I was just wrapping up a three-year degree, and now I was starting my first professional job in the field.

When I was in university, I had no idea what to expect in the field of software engineering. I didn't know what domain I wanted to be in or if I would enjoy it. Based on my past failings, my biggest concern was being awful at it. University did a poor job of preparing me for the real world; I was left with many unanswered questions.

In this book, I will attempt to answer as many of these questions as possible and answer the ones you haven't thought to ask yet. I will discuss my experience during my first year as a Software Engineer—what went well, what went badly, and what I learned from it all. Furthermore, I'll discuss the various lessons I learnt over the year. These will give you a jump start on your competition. I will incorporate the advice I have gained from others over the years and advice I wished I had had when I started. I hope that by the end of this book, you will understand more about the field and whether you want to enter it. After you read this book, you should be thoroughly prepared for your first year as a Software Engineer and the long, successful career you can have.

SECTION 1

My Experience

MY FIRST YEAR

I started my first job at a consulting company in Melbourne, Australia, in January 2018, and the year flew by. Consulting companies differ from regular companies, like a company that produces X bit of software, for example, Spotify. In a regular company, you could be put in a team that focuses on a small part of the application, or you could be put on some new product they wish to make. However, in a consulting company, you could be put anywhere on any type of work at any stage of the project. You could fix bugs in a Python API one minute, then build a new product with a NodeJS back-end. When you work for a consulting company, you could be on a project for one week or one year; it varies from client to client, project to project, and company to company. My first project was to work on a start-up in the education space. It was a brand-new product, and luckily, I got to be a part of each stage of development. I was lucky enough that my first project was for a start-up. I was fortunate because not many junior developers get to do anything more than bug-fixing some legacy applications. It only took us six months to get to production, which was quite a remarkable achievement. I spent the first month working by myself for the most part, but also with my boss to get to grips with the programming language Elixir. Having learnt object-oriented languages at university, I was stunned at how different the language was. Not only was I changing the paradigm, but I was also moving from object-oriented to functional (the difference between the two, I could not tell you at the time), and I was starting a brand-new job right out of university. I struggled with the beginning of this project, having a solid freak-out every day or so. My boss told me to buckle down and get going, so I did, and with some help from a book, I managed to get some form of an API working with GraphQL. I even had the idea to try out a graph database, which was fascinating. Unfortunately, we struggled mightily with getting it deployed to any type of server, and the library interfacing with the database and Elixir was extremely new. My boss decided it would be best if we changed back to a regular relational database, and he was right, even in hindsight.

After three months, my boss decided we needed some help; otherwise, we wouldn't hit our deadline, so the company contracted two back-end developers with Elixir experience, one from Poland and one from Sydney, Australia. At first, I was annoyed; someone would be taking over my codebase and taking credit when I thought I could get it over the line myself. I couldn't have been further from the truth; these two developers were well-experienced in both the language and the field, and as a result, they helped me grow 100 times faster than if I had done it on my own. They also provided more general development experience to the client, experience that I would end up learning from over the course of the year.

After six months of hard work, we finally produced the application and released it to the market. Over the following six months, I learnt a metric tonne about handling production, deployments, migrations, existing datasets and angry customers or, more to the point, an angry client with an angry customer. All these experiences aren't something you necessarily get as a junior, but I was fortunate in that I was a part of many of the different scenarios you might face over a career in some form or another. I experienced a client that didn't believe in me, and I broke a production database. I dealt with clients who didn't listen to junior developers, and I worked with remote developers. Most of all, I worked on a start-up with a new language, with great people, and I sore the result of my labour used by people.

At various points during the work for the start-up, I had to fill the lead developer role whenever my boss was busy or had to work with other clients. I observed how my boss handled the clients and how the more senior developers handled themselves. I took this chance to step up, leading the team when I could, and I loved it so much that I asked my boss to lead another project.

About nine months into my first year, a project for the Starlight Foundation, a charity in Australia, started. I was asked to be the lead developer in building the mobile application they were after. I was part of a team of two developers, one project manager and one UX designer, and it was the perfect start to my leadership career. They were after a mobile application that could consolidate several tasks that a staff member would perform, such as timesheet, availabilities, chat, expenses, and single sign-on between the previous features and whatever external system they used for that task. The single sign-on feature was one of their biggest concerns; it would reduce friction in how the staff members operate, allowing them to spend more time helping the children the charity was built for. The project manager on our side had no technical knowledge, and the project manager on Starlight's side also had no technical knowledge. I was brought on to be the bridge between the technical and non-technical sides and to code whatever needed to be coded. After we gathered all the requirements, I brought multiple developers together to test the SSO idea. It was determined that it would be next to impossible, especially given that we lacked control over one of the applications. If you don't know what single sign-on is, it's the feature that allows you to login into one application, and when it kicks you to another, you are already logged in on that one as well. The perfect example is the Facebook app, where you log in to the main application, but when you want to chat, you get sent to another app. You don't log in again because Facebook owns both applications, and when they send you from one to another, they also pass along your credentials, so moving between the two becomes seamless. SSO is a fantastic feature, but it becomes impossible when you don't own the other applications, as we were about to find out. I arranged for a week-long deep dive into how we could/if we could achieve single sign-on and to what external applications. They already used Microsoft services, namely active directory, meaning if an app supported it, we might be able to use active directory to handle the login. This allowed us to SSO into a few services, but they were only minor and insufficient to warrant a new application. The next step was hooking it up to external applications. This failed miserably, as we expected, since we didn't own the other applications, nor would we be able to send the credentials in any way. After a week's worth of work, four developers and a lot of trial and error, we concluded that it would be impossible for their needs and scenario. I now had to explain to Starlight's project manager why it would be impossible to build what they wanted. It wasn't as hard as I expected to explain to a non-technical person why we couldn't do something highly technical. It gave me my first experience handling a non-technical client with technical problems. Furthermore, it was unlike the start-up, where the client was, in fact, quite technically savvy. Unfortunately, the project was scrapped due to technical constraints as these were too important to the client, and without them, it left them with a mobile application that wasn't worth building. It wouldn't solve any problems; it would probably create more.

I also managed to meet one of my best friends in the first year of work, which, thinking back on it, is quite impressive and most likely won't happen again. However, at each job, I have met great people that I still talk to today. You meet plenty of people; some of them you gel with, others you don't. It's all part of the experience.

Coming from university, which is quite relaxed here in Australia, I found it quite hard to transition to full-time work. In fact, one time, I fell asleep while my boss was talking to me. I was just that tired from jumping into full-time work. Every Friday, I would stumble home, tired, and spend Saturday and Sunday recovering, all to do it again. I loved my job, but the physical toll of going into an office every day is thankfully a thing of the past. If you find yourself struggling with the demands of an office, working from home is a game changer. You learn all about these little things you never put a thought into. You find the things you like and dislike in software engineering. It's a wild ride but one that brings forth a lot to your life. Enjoy your first year because it's a blast and goes by quicker than you can possibly imagine.

This sums up my first professional year as a software developer. It was one hell of a ride, and I wouldn't have changed anything. It was tough, but I learnt an incredible amount, much more than I thought I would in my first year. Your first year is easily going to be the hardest. Everything is new, and everything is hard, but it's the hardest days that push you to be better. The hardest days create the greatest growth. When you reflect on where you have come from after the first 12 months, you won't recognise the person in the mirror.

THE GOOD

The first year of my career was fantastic. Sure, I had some challenging moments. I broke some production data and freaked out a lot. I also struggled to get up to speed on a new programming paradigm and language. Yet, through it all, I managed to appear from the other side, knowing vastly more than I did when I began. I wouldn't be anywhere if it wasn't for the surrounding team, the more experienced developers that helped me along the way, the great front-end developers I worked with and the incredible boss who told me I could do it when I doubted myself. The major key to the success of my first year was working with amazing people. I made remarkable friends; one of them even became a groomsman at my wedding!

The biggest thing I learnt from university was that what you learn in university only somewhat prepares you for the job. You must understand you don't know anything. You will be floundering around, not knowing which direction is up, but you don't have to worry because companies expect you won't know much. They will either put you with more senior people to learn faster or give you tasks such as bug fixes. I stressed big time before I started that I wouldn't know anything and would be fired within the week. You won't know anything; the quicker you realise that, the easier it will become. There's nothing worse than a junior who thinks they're remarkable when they don't know much and refuse to take advice from others. You won't last long in the industry if you don't take advice from those around you. The biggest thing I learnt was to swallow my pride, put aside my ego and realise that I can learn something from everyone no matter their skill level. I have been blown away time and time again by people with minimal experience teaching me something entirely new or showing me a way to think about a problem differently or a language feature I didn't know existed. Everyone has a different perspective on life, and with different experiences, you can learn a lot from anyone; it's part of the spice of life.

Working with Clients

I was grateful to get the opportunity to work as the 'lead' developer on a project for about a week. This isn't something you normally expect to do in your first year, but I did, and I learnt a lot about managing people, even if the project was scrapped early due to technical constraints. However, I still learnt a lot about handling a client and how to explain to a non-technical client why something would be impossible, even if they didn't understand it at first. It requires significant effort to communicate an understanding of a problem to a client.

Client management is one of the hardest things in the field. You might work at a product company and never need to manage an external client, but you still have to manage someone and their expectations. Expectations are some of the hardest parts of software development. You can't control how others feel, but you can taper expectations with solid communication skills. I would encourage anyone who works in the field to work at a consultancy/agency for a year. You learn a whole new side of yourself. You build communication skills that you simply do not get from working in product companies. Furthermore, you have to manage the client in a way that keeps the relationship going and them happily paying for your services. In a way, you are constantly proving to them that you are worth their hard-earned cash. This gets easier and easier as the relationship builds, but the communication you bring to the table builds this relationship. Communication is everything!

I'm most proud of what I achieved in just one short year and my first year at that. I was given an opportunity multiple times, and I took it and ran with it. I learnt more than I ever thought possible, vastly more than my entire degree. Likewise, I learnt that what you learn in university is probably going to be out of date, so don't worry if you only just pass. In fact, I would put more emphasis on getting an internship than getting a degree. The experience, knowledge, and real-life programming you perform are infinitely more valuable than going to university. Sure, you get a piece of paper at the end of it; however, depending on where you're from, that could cost anywhere from free to hundreds of thousands of dollars. It simply isn't worth it if you have to pay it back until the day you retire. If you can find somewhere to intern, then that is your best bet. You might even get paid for it, at least here in Australia. Even if you don't get paid for it, there's a good chance you're quite young and living with your parents, so you might be able to take 3-6 months of unpaid work. If you're not in that position but have savings, you could also live off that. If you can't do any of that, you might live in a country with some form of social security payment. There's a good chance you could live off these, which would be similar to going to university, except you might have a paid part-time job there. It's all about the experience you gain. Perhaps university is for you, so take full advantage of it and build things in your spare time. Or spend the time you could be studying for high marks on something you could show off to an employer. Walking into a job interview with a fully functioning website is vastly more impressive than walking in with top grades. The top grades will be in concepts you will most likely never use on the job. Building an entire website for a company that builds websites is the direct experience that points to you being competent and hireable.

If you ever get an opportunity to take an internship as a programmer when you're young, you should take it, even if it's not paid. The experience you gain from this is immeasurably valuable. I wish I had the opportunity myself. I remember when I was 16, and the school made everyone work one week in an internship. The majority worked at chain big box stores, but I was arrogant and wouldn't look, nor would I take any internship that wasn't paid. I didn't do one, so I stayed home for a week instead. However, later, when I tried to get a job, it was arduous. I had no work experience. It took me ages to find a job, but it would have been much easier if I had the internship under my belt. The same goes for a career in programming, yet it's even more important. When you're a teenager, there's a good chance you have no idea what you want to do, and that's okay; I didn't, either. However, if you do, please apply for internships or find someone you know who might give you one because they're worth their weight in gold.

Picture this, you're 21, and you just finished your degree. You're hungry and looking for a job. You have no work experience except for working at the local McDonald's. You have a massively overpriced piece of paper in one hand and a caffeine-induced shaking right hand from the all-nighters you had to pull to get that piece of paper. You apply for hundreds of jobs and get next to no responses. What do you do? Well, there is not much you can do aside from continuing to apply for jobs and hoping for the best. It might take over a year to find a job, but you're set once you finally get one.

Now picture this, you're 22, and you just finished your degree. You're hungry and looking for a job. You have six months of work experience and time spent working at the local McDonald's. You have a massively overpriced piece of paper in one hand and a caffeine-induced shaking right hand from the all-nighters you had to pull to get that piece of paper. You performed an internship for 12 months via the university you attended, and as a result, you finished your degree a year later. You walk straight into the job you worked at for your internship because you put in the effort and weren't a complete idiot. This frequently happens. Universities offer countless internship programs, but if they don't, you might know someone, or if you're really lucky, you could do what I did, but this was a coincidence.

How I Got My First Software Experience

I was working my shifts at the local Subway chain. The more you work, the more you get to know the regulars. This group of three men would come in every day and order the same thing every time. On one occasion, I overheard them talking about some programming-related issue, so I took this opportunity to ask them about themselves and found out they were, in fact, programmers. I talked to them for a bit and then introduced myself as a university student looking for internships. They said we would love to have an intern, and that's how I got my first internship. This won't happen to everyone, but taking these opportunities leads to much greater things. In fact, all my jobs have come from knowing someone or talking my way in. My second job was through the boss of the place I got the internship. My third job was from a friend at university, who recommended me to a company his current company was working with. It's more about the people you know than what you know. James, from my time failing subjects at Monash University, was someone I kept in touch with throughout my degree. When we needed to expand the engineering practice at my first job, I thought James would be the perfect fit. I knew James was about to finish his degree, so I asked him if he would be interested in an internship. It just so happened that the place he was currently interning at had a hiring freeze, and they weren't going to extend him past his 12-month internship, so he took the offer.

These situations continually happen. Networking is one of the most important aspects of job hunting. The best part of my university degree was the networking I did, and I am still friends with three people from my degree, one of whom got me my first full-time job post-university.

Learning

Any good/established company will have a program for mentoring juniors and graduates. This process is critical, but I didn't get that at first since I joined a small consultancy after graduating from university. However, I did get it eventually. When we got our first big project, I was chucked on that with no real help, but it was alright because it was an internal project. I struggled until we hired the developers from Sydney and Poland. They became my mentors. They taught me right from wrong, reviewed my code, taught me best practices, and, most importantly, they were patient with me. This is how you want your first year to be. You need to be taught and be around more experienced developers as much as possible. That's the only way you're going to learn. If you're always working by yourself, you continue to perform the same way and never get pulled up on anything. One memory that sticks out is when I created a merge request for a piece of work I had completed. The two senior developers ended up writing over 100 comments on the pull request. It was hard, but it made me a significantly better developer. It took hours to go back and forth fixing issues here and there, but I'm glad they did it because I learnt a lot.

I had multiple internship opportunities, the first two lasted about a month or so each, but the third was with a consultancy for three months, for which I later became a full-time employee. This last internship was the best. It mostly consisted of me being asked to research or build a prototype of my boss's idea. I spent two weeks working on a blockchain proof of concept and then a few hours thinking of possible ideas that could be done with blockchain.

I heavily relied on learning sites to watch video tutorials to build enough knowledge to create something useful. Furthermore, I watched everything and anything from cloud-related tech to front-end frameworks. I spent my internship learning a bunch of stuff, broadening my knowledge. University mainly got me to focus on back-end technologies and some theory, but nothing related to the industry. It wasn't until I got into my first jobs that I truly began to learn about the industry.

The greatest thing I achieved in my first year was going from knowing nothing about the programming language Elixir to having real customers using the Elixir/React application. This isn't something many junior developers get to do, and I'm truly grateful for it. Over the course of a year, we took a start-up idea from pitch to production with real customers. It was both a testament to the technologies we used and the team that did it. I still can't believe how much a small team produced in one year!

THE CHALLENGES OF MY FIRST YEAR

Growing Pains

It's challenging to start a development career, or any career for that matter. You know absolutely nothing when it comes to development when you start. I didn't even know how Git worked or how to properly push code when I started. I had used Git at university for a subject, but it made little sense. That was more of a reflection of the teaching than anything. One of the funniest examples of knowing nothing was that I went into back-end development because I was never told that as a front-end developer, you don't have to design it; you get designs! Not only do you get designs, but there are also people who specialise in design, and funnily enough, they're called designers. Somehow not a single soul at university mentioned this, and I wasn't alone. Two friends who I attended university with also didn't know. It wasn't until we got into the field that there were designers and front-end developers, neither of which did the other's job. Sometimes I would see some code or pick up a ticket and freak out, not knowing what to do or how to start.

During my fourth week on the job, I freaked out because my boss asked me to write a bit of code in two different languages and report on which one was easier. I was stressing big time. I didn't know either language; I didn't know much of anything, to be quite honest. My boss told me to slow down, so we sat together, pairing on the problem until it was done. I learnt a lot in my first gig, mostly from pairing. At times, it was frustrating when I tried things on my own but needed to talk to someone for help at some point or another. Honestly, that's just being a developer. Even now, I still need help on things, whether it's some part of the codebase that I've never worked on before or some weird behaviour that makes no sense at the time.

As you get better and better, you will need to ask for help less often, but it will never go away; you will never know everything about a codebase or programming. I still regularly ask for help, less because I don't know something and more because I would rather not waste time floundering around, especially when there is probably someone who might know that area better. Most projects have tight deadlines, or you might feel embarrassed that you don't know something. At the end of the day, your team and boss will be much happier if you can save hours by asking one question than looking like a hero who can solve everything. The mark of a good programmer is how early they ask someone else and how well they know their strengths and weaknesses. For example, one of the developers in my team spent an entire day debugging something with some weird behaviour that, in reality, made no sense for what they were working on. I could debug and answer the question in about one minute because I had discovered the bug earlier that day. Even when it's an issue that I hadn't already discovered, having more experience allows me to suggest or work out things much quicker, which can save a tremendous amount of time. When I'm working on a codebase with many developers and in an area that I'm not familiar with, I'll ask people immediately. When odd things start happening, I'll debug for a bit, and if I'm still stumped, I'll grab a second pair of eyes. If those fresh eyes can't solve it, I'll deep dive. You might feel like you are annoying others by asking for help, but at the end of the day, if they're annoyed, it's their problem; you are doing the right thing.

Lack of Help Because I Was at a Start-up

I was working at a small consultancy on a start-up as my first client. This meant that I was often my company's only back-end developer on the project. There was another more senior developer to guide me at times, but if he wasn't there or working on something else, which was frequent because he was from a different company, it was up to me to figure things out. Sometimes, he just wouldn't know how to do something, so again, it was up to me to figure it out. This is a common difficulty when working at a start-up or for a small company; there is a lack of resources to help you, which is especially important early in your career. The lack of resources can also lead to some bad programming habits. You must always keep in mind the habits you are building and not stick to something just because you have been doing it the same way. It helps to try to solve problems from a different approach, to hopefully learn a different method or feature in a language. With the lack of developers comes the lack of reviews on your code. This, again, can lead to poor code quality and a lack of learning. You must stick to learning and improving yourself out of hours and during the 9-5 so that the lack of resources doesn't come back to bite you when you go to your next job, and your code quality is awful. You will get excellent at googling and reading documentation. It also helps that this is one of the most important skills for being a developer. As a result, you will progress much further in this skill than most developers of your experience level. On the flip side, you have much higher quality code if you work in the enterprise space because many people are reviewing it. There are examples of how to test things and build things in certain ways. For the most part, in an enterprise, your problems will have been solved by someone else, somewhere else, which is fantastic, but it can also stagnate your growth.

Mistakes

I made many mistakes, and so will you. In fact, I still make plenty of mistakes; it's unavoidable. You won't ever grow out of it; if you are still making mistakes, it means you still have more to learn. You make mistakes every day for your entire career. The key is to learn from them, grow and not make them again. If you aren't making mistakes, it's not because you're a remarkable developer; it's because you aren't being challenged enough, and you won't be growing. Growth, in my mind, is directly tied to how many mistakes you make, so if you are constantly making mistakes, you are continuously learning and growing. Just be certain not to make the same mistake twice. You learn the most from straddling the line between hard and impossible. This line pushes you to become better.

You will look back on the code you wrote just a few months prior and think, *wow, that is some horrid code.* Everyone feels that way; it's just a sign of how far you have come. You will see the mistakes in your code or the lack of tests. You will see the complexity of the code that could have now been written much more simpler. This is all part of the learning process. The more mistakes you make, the more code you write, and the more time you spend reading other people's code, the better you will get. You will often catch the mistakes you've made in the past, which pushes you to become a better developer.

Thrown in the Deep End

Depending on your personality, this could be a good or bad thing. I loved it in numerous ways, but certainly not at the time. Being thrown into the deep end is horrible; you feel like you're constantly drowning, and really you are. It's not until months later, when you reflect on your experience, that you realise how great it was that you were treated the way you were. When you look back on how far you've come, you see how you went from someone who knew absolutely nothing to someone who knew a little more. I love being thrown in the deep end because you learn vast amounts in a small period of time. It's the best way to learn, but it's not for everyone. If this is not your preferred way of learning, look for jobs in larger enterprise companies. At an enterprise, everything moves slowly, and there are many people. There is always someone who can answer your questions. The pressures of deadlines are often more relaxed or open to pushbacks. At the very least, as a junior developer, you won't feel them; that's for your boss's boss to worry about. This is quite the opposite in a start-up or a small consultancy/agency where I started. As a junior developer, you will feel all the pressures directly, and there will be next to no one to ask questions. You will spend a lot more time wondering what to do or how to solve a problem, but in this wondering time, you will understand a lot more about the software you are building and the project. There is no better feeling than the triumph you get from solving a problem you have been stuck on for days. One of the best feelings as a software engineer is the feeling of solving a bug or finishing a ticket that has given you grief for days. You're on top of the world; it's a feeling we wish we could bottle. Some days I think this is a feeling that all developers chase relentlessly.

Burnout

I went into my career fast and with reckless abandon. I got in early, stayed late, and spent every second I could learning something new. Furthermore, I prided myself on being the first in the office and the last to leave. After 10 months of this, I was pretty burnt out. I struggled through my day, and I took longer on tickets. You will probably get burnt out during your career, most likely in your first year. I certainly did. I had to take a big step back and only work eight hours a day and nothing on the weekends. If you burn yourself out, it will take a few months to recover, and you might want to consider a holiday. Don't be like me and burn yourself out with your excitement and ego to be the best. You will get there; you just have to be patient.

Burnout will destroy all your motivation and even cause you to hate your job if you don't listen to the warning signs. You simply can't work extensive hours for long periods of time. You will quickly learn how long you can work each day, how many days you can work, and over what time frame you can keep it up. Furthermore, you are not Elon Musk; you can't work 20 hours a day. You will be lucky to produce more than 4 hours of code a day. Your brain is surprisingly power-hungry, so when the batteries are drained, you have to give it a rest. It's the same for the long term. Burnout will be apparent. You will begin to feel more tired and get more frustrated. You will get more annoyed at the project when a few weeks ago you loved it. Likewise, your level of motivation will be so low that you will hate your job. This is all burnout, and it quickly saps all your productivity if you don't address it. If you catch it early enough, you can quickly recover with a long weekend or a short holiday. Be warned, if you wait too long, you will need months to recover. You require hobbies to stop you from working all the time. Hobbies are the perfect way to unwind, relax and follow an interest.

This sums up the advice from my personal experiences over the first year as a developer, and yes, I really did fall asleep when my boss was talking to me!

THE UGLY

The Not So Great

I caused my company to lose a client on my first project. I was working on a codebase with another company, and we were somewhat competing for the work but working together nonetheless. We brought this company on to help with the development as they had more speciality in the language, and at the time, our company only had two developers, one front-end and one back-end (myself). This meant that if we wanted to get this released, we would have to get some help. We partnered with a company from another state in Australia, and they allocated one extra developer in the back-end who had a specialty in the language we were using at the time, Elixir. We worked together for 10 months, and I was instructed to ask as many questions as possible. This was my first full-time development job, so I took that opportunity to ask plenty of questions.

After the 10 months were up, we released it to production, and several clients enjoyed the software. It became time to start thinking about the next phases of the client's business, for example, where would they release next locality-wise and what features would they add? The client approached us to say they would no longer be working with us, instead continuing with the other company. I found this a bit strange at the time, and it wasn't until almost 12 months later that my boss told me they had left because of me. I had been too junior and asked so many questions that the other developer convinced the client that I was sucking up all his time and unable to do anything as a result. This meant that I was directly responsible for losing the client. At the time, this was devastating because I liked the developer I worked with and was under the impression we got along well. But being stabbed in the back was quite painful. I later took this as a lesson: retaining a client is not part of a junior developer's job description. The team members are there to help you out just as much as you are to help them out. However, you always need to be wary of relationships. Most people are not like this, but some will do anything to get ahead. These situations happen, but don't take them personally. Move on and always act professionally. You shouldn't take advantage of others because the toll it will take on your soul isn't worth the extra money in your bank account. You don't have to sacrifice your ethics to get ahead. Furthermore, you can build others up, work with others and be proud of their achievements because there is enough for everyone.

Breaking Production

I broke production while writing some database migrations to edit a few tables. Database migration is when you change fields in a database table, and you have to write a special script to handle the changes. It's not hard; however, you need to be cautious. I wrote my migration script to tell the database exactly what I wanted to be changed and how I wanted it changed. I tested it locally, and it worked perfectly. Then I pushed my code to production, where it promptly failed and failed to roll back. When a script fails, it attempts to revert what it did, much like when you undo in Word. I hadn't tested my rollback script, and it promptly failed and put the database into an obscure state. I also hadn't tested my migration script on some form of production like data, and it failed, and thus the database was in a weird state and completely down. It didn't take long for the sense of panic to set in. Real users were trying to do very real things at that moment. I quickly got on the phone with the other developer, who guided me through what I needed to do. I was able to find out what the error was. I promptly wrote a script to fix it and re-migrate the data, and it was back online again. This situation was terrifying, but it will happen. From then on, I was extraordinarily careful when writing anything that would affect the database. I tested it extensively, ensured it had realistic data for those tests, and then watched it like a hawk when I put it into production. These are the lessons you will learn a lot from; you can't get this experience any other way. You will freak out, and that's normal, but take a breath, clear your mind, step through the problem then fix it. You learn something from the process, and hopefully, you never repeat it. There is a chance you might get yelled at (you shouldn't), but that's the life of a software developer. Everyone will break something in production at some point or another; it's just about when. Do it early, fix it promptly, and don't freak out.

Clients

I mentioned my story of losing a client, but you will have to deal with clients in any consultancy or agency, which is great if that's your jam. Unlike in a product company (a company that produces a product, think Atlassian), you get to be a great developer and communicator. You must listen to the client, work with them, put up with them, and handle them. You learn a lot about people when pressures, time, and egos are involved. Furthermore, you will screw things up, sway their opinions on some things, and more than likely, they will ask you to produce something that is downright insane or impossible. This all leads back to communication, how you portray yourself and how confident you are with your knowledge and your team. You might be unable to sway them to every decision, and occasionally, you just have to do what the client wants. However, you have done your job if you can stand there and give them a clear and well-explained reason why they can't or shouldn't take a particular path. This can often be the hardest aspect of your career; just take it as a learning experience each time, and you will get better and better.

Interesting Characters

Some people are just odd; they might not talk to anyone, think they are the best thing ever, or push their views down your throat like a mother bird feeding its young. No matter what, they are people all the same, and people can be the hardest part of the job. People are also the most important part of any company and any relationship; thus, if you can't handle people, you can't handle software engineering. In my mind, you will be challenged more by people than by the code. Code is easy; you get there after a long enough time and enough pain because computers are predictable. People can act irrationally; they can take things the wrong way and burn a bridge you built for months. People can be tough to deal with, but it's worth it because when you work with great people, you enjoy your work magnitudes more. If you can handle people, you can handle software engineering.

Some people will play games or put others down to get further ahead, while others will stab you in the back, given the slightest chance. Yet most of this is rare; it's more about the personalities you must deal with that are the most challenging. You need empathy for people; having empathy is the start of a good life. If Bob rewrites your code and tells you how bad you are at programming, you could take that extremely personally. Now, this is never okay, but if you were to reframe this with empathy instead, you would notice that they lash out because of some underlying issue that has nothing to do with you. Once you see others through an empathetic lens, you begin to feel happier and no longer take things to heart.

Take responsibility for how you perceive events, as this is the only thing you can control. Approaching others with empathy will change how you see people and situations! You spend 40 hours a week with your colleagues, so you're bound to see their bad sides. If you ask them if they are okay and approach them with empathy, you might just find out they are struggling in their personal lives.

Don't take things personally; it's seldom personal. If someone does something to harm your career, it's more for them to get ahead. You can't control it or change it. Accept it, move on and enjoy where you are. The more you can appreciate your situation, the happier you become and the more enjoyment you get from work. Things bother you less when you cast them aside. You can't simply delete your emotions, but you can reframe your situation. You can tell yourself that what Bob said won't bother you since he was upset because his boss was putting pressure on him or something was happening in his personal life. There could be 1000 and 1 reasons for comments, and rarely is that person out to get you. Casting aside these situations is important because the more you dwell on them, the more they affect you. Each time you think about what Bob said, your body relives that experience. Your body can't tell that it's not happening again; it's being attacked by something Bob said weeks ago, and the same stress response is being triggered. Reliving these memories is reliving the trauma. However, casting these experiences aside is the opposite; it frees your body from the stressor. You didn't have the choice to get verbally abused by Bob, but you don't have to relive it. That is your choice.

WHAT I EXPECTED VS WHAT HAPPENED

I didn't know what to expect coming into my first year. I had done a couple of small contract jobs throughout my time at university, but nothing major. In fact, the first contract gig I got, I was let go, which made me even more nervous about starting my first full-time job. My biggest issue with previous work I had done was that I never had any help, which, as a junior, is important. You need someone to guide you, answer all your questions and tell you that you're not dumb, but this is tremendously hard to come by. I was ecstatic to begin work; however, as I said, I was quite nervous about it at the same time. My fears were quickly squashed when my boss would routinely help me when I ran into problems and my colleagues routinely reviewed my code, which allowed me to grow exponentially. They would tear my PR apart (nicely), telling me what could be improved when I wasn't using a technique properly or how to implement something much easier. This constant feedback on my work was instrumental in my growth, and this will be the cornerstone of your first year.

Growth

Your first year is about growth, improving, soaking in the knowledge around you and building the good habits to have a successful career. For the most part, you're habit-less, but soon enough, you build strong habits around how you code, the style you use and how you solve problems. The people around you influence you most; they shape you by reviewing your code and imposing their coding style, which you will transfer to your style. You learn what good and poor code is, how to follow the project's conventions and how to work in a team. Most of all, you grow, expand, and flourish. That is what the early part of your career is all about—growing rapidly, getting opinions on your code, problem-solving and refining your problem-solving and delivering better quality each week. Your goal in this period is to be humble and realise that people are not attacking you. They make these comments to improve your code and the code of the entire team. You begin knowing nothing, and even if you think you know things, you don't; you have to let go of what you think you know and let the team and project guide and teach you. Of course, you have opinions, but generally, the more senior developers have more thought-out and experienced opinions, so take them on board as much as possible.

Try to soak up as much information from the most senior people in the team. Reach out to them regularly for advice to review your code and help solve a problem if you are stuck. Don't wait too long to get unstuck; reach out early and often. These are all things I never thought about when I started, as these things are not taught. You learn more from the people around you than from completing tickets. Sure, you learn a lot from doing, but soon enough, you solve the problems in the same way. Over time, you receive less value from working alone until you only solve every issue one way. The real value comes from learning with others and from others. No person in the team is without knowledge. Everyone has different perspectives and experiences, so use them to your advantage.

If you regularly go into the office, spend time chatting with your peers and get to know them because it's the best way to learn. People are fascinating when given a chance.

Projects

If you work for a product company, you most likely get placed in a team to deliver a software project or in a team that performs maintenance or bug-fixing work on a piece of software. When you work for a consultancy or agency, there is a good chance you are building something from scratch or being put in a team to finish the delivery of a project.

Enterprise Development

Most people will work at a medium to large-sized company, which is probably the best place to start. There are more guide rails, more people to work with, and more processes and established practices to go off. Most problems in enterprises are solved by someone else, and if they aren't, there is some genius who can do it in five minutes. All this is great for your development as you can ask, work with and absorb all this information. You have opportunities to move teams at regular intervals, meaning if you get bored with API work, you can switch to the front-end or vice versa. There are clear grade structures with clear objectives to hit, which means you know exactly what you need to get promoted. This will all be explained in the first couple of months. The flip side is that you could potentially be placed into a team that doesn't do much of anything, and if you are like me, this will drive you crazy with boredom. You might find that after your first year, you have learnt very little, depending on the project you are placed in and the decade that project was started! Being placed into a maintenance team of a core bit of the business could be the slow death of you because you will be resolving small issues here and there in a codebase that is older than you.

Consultancy/Agency/Start-up

If you go to a consultancy or an agency, you will have a diverse time and skillset. More than likely, you build a new piece of software, which is great because it means working with the latest and greatest tools and technology. You will be on the bleeding edge, doing cool and interesting things and learning technologies that will be hot and in demand. This is one of the biggest benefits of a smaller company. Often, you work with modern technologies, widely used in the industry and highly applicable for future employment.

In a larger company, depending on their age, you could find yourself stuck on a legacy system with very few future job prospects. This is not to say that this will happen; it is just something to watch for. The work you do day to day needs to also contribute to your resume. If you find yourself learning an extraordinary amount, keep it up. If you are going nowhere, then move on, but make sure you have a foundation of transferable skills to do it with. Later in your career, it won't matter what your foundation of skills is as you will pick up most technologies rapidly. However, when recruiting junior developers, the less time you spend teaching them the skills for the job, the easier it is for the team. Therefore, one year of developing a specific skill is important for a junior, and why it can be hard to move to another technology early on. Once you progress in your career, it won't matter that you haven't used a specific technology; you will have used something similar, and if you haven't, you can figure it out.

I almost fell into this trap myself. My first year was spent using a language called Elixir. It was a fantastic language; at the time, it was quite new and not used by many places. In fact, unless I boarded a plane for 16 hours to the US, I was hard-pressed to find any companies in my country that used it. Furthermore, it was also a functional language, which wasn't particularly popular at the time. After my first year, I looked around at other jobs, but finding one using that language was almost impossible. Every job I applied for would reject me because I didn't have the right technologies at the time. I stopped applying and spent the next year learning everything I could about the technologies that all these jobs relied on. I created small side projects, I watched tutorials, and I read. After the year ended, I might not have had the year of work experience, but I did have the knowledge to talk about them. I could fudge the numbers a little on my resume, claiming I had a year when obviously, I did not. However, the main difference was that I could talk about these technologies. I wasn't lying per se, but it worked all the same.

The downside to smaller companies is that they are typically a lot less established. There are fewer guide rails, fewer people around to help, and more pressure for deadlines to be hit. This can either be a good or bad thing, depending on the company and the type of person you are. If you dislike the pressure, focus on an enterprise company since you will be effectively shielded from deadlines. If you are eager to learn the most, then a small company will allow you to wear many hats by necessity and perform many tasks, pushing you faster and harder than you could imagine. You will finish your first year knowing an incredible amount and having the ability to show off a considerable range of skills to new employers. You may not have 10 years of experience in SQL, but you could at least speak to how you construct a database and how you would normalise these tables.

Furthermore, given the nature of the company and its constant looming deadlines, you will pick up some bad habits. If you keep your head on straight, output the best code you can and band together with your teammates, you can have effective code reviews and grow as a unit. It's even better when you can pair program with your teammates, as it takes you and the other developers to the next level. Furthermore, you get to discuss your experiences and share a bond. Pair programming becomes a great way to bond as a team, one pair at a time.

What You Won't Be after a Year in the Field

You will still know nothing about many things. Your confidence will be at a 10, which is way higher than it should be, but that is great; confidence is where you want to be. The Dunning-Kruger effect will be in full effect, meaning you will have all the confidence in the world but lack the skills to back it up. You think you're a senior-level developer already when in reality, you're just a much better developer than a year ago. This is great in some sense, as long you don't let it consume you. You won't be as great as you feel, so you must stay humble. Hopefully, you will have developed a great base of skills and knowledge and can now be left to your own devices when tackling a ticket. However, you will still need plenty of help since you will have limited experience in various technologies, but you will get there. You will need to approach the second year as you did the first. Soak up the knowledge, stay humble and don't get overly confident. If you're lucky, you might have been a part of the team that delivered a project, but if you weren't, that's okay too; not everyone gets to deliver software to end users each year. You will still make many mistakes and require extensive code reviews, but there will be fewer comments on your code. You won't have the skills to design a solution to a client's problem or create the next Facebook, but you will have better skills and a hunger to improve. If you come out of the first year with nothing but a hunger to learn, it's been a successful year, and you should be proud of yourself.

One year is a minuscule amount of time. You will have spectacular highs and very low lows. It's conceivable you will deliver software to a user; it's also probable, if not certain, that you will push a bug into production, which causes issues for a user. These are all learning experiences, and you should take them as such. Grow from every little mistake you make, and stay humble for year two. Your career is a long one. Taking it year by year, month by month, week by week and day by day will ensure you have a successful one. Don't become complacent after your first year; continue to do what drove you to success, and it will pay off in enormous quantities in the future.

YOU CAN ONLY CONTROL YOURSELF

In life, we stress about the problems we can't control. We think about the person who pushed in line at the supermarket or the hail storm that damaged our car. All these difficulties we have are real. They hurt, drive you mad, and make you sad, but you can't control them. You can't control how Bob feels about you or solve the world's problems. But you have the power to control yourself and yourself alone. Life might be out of control, but you can remain in control of your being.

For example, if you complete something you're proud of at work, but Bob comes along to review it and tears it apart, your confidence can be destroyed, especially if it happens repeatedly. However, you can decide to use this as a learning opportunity. You can ask Bob to explain his comments and ask what he would have done to improve. If he's just ranting, get someone else to review your work. However, if his points are valid, address them and improve.

I'll say it again; you can only control what you can control. Everything else is out of your control, and you should put it aside; otherwise, you will drive yourself mad. You can choose how things affect you, but you can't choose how others react to you. The hardest part of your career will be dealing with people before any technology. Some people won't like you, and some will just be rude. You can't control them, and you never will. Therefore, you must learn to control yourself and that with which you have immediate control.

SECTION 2

Career

INTERVIEWING

The Process

All companies have slightly different interview processes. If you're a big company like Atlassian, you might have a 12-step interview process that can take months. If you are a smaller company, you will probably have around 3-5 steps. I am a big fan of the three steps; the fewer steps, the less time I waste if I don't make it through. However, a typical recruitment pipeline will work as follows:

Resume submitted

The most interesting ones get offered an initial interview.

Initial interview

Have a 30-60-minute call about what you are looking for and why you applied for the job, and briefly talk about your experience level, background, degree, achievements, and what you want in your career.

Take-home test

(More common) This is a test where you must solve a problem to prove you can code.

In-person test

This involves a live coding session in front of a senior developer to assess your skill level. This is usually easier than a take-home test.

Take-home test and in-person test

Some companies do both, where you complete a bigger take-home test and either talk through your solution or extend it live. This typically means the take-home test is rather straightforward, and the live extension will be relatively easy too. However, if you are stressed about programming live as I am, you will find it hard regardless of what path you choose.

Final interview

Typically, a final interview includes the live test extension/talking through your take-home test and Q&A sections. The Q&A for a graduate/junior position is primarily around Git and Agile, and how you work and work in a team. If you have had previous experience, you will be prompted to draw from that. These are all relatively easy questions. However, some companies might ask about algorithms, data structures or equally complex questions, but it looks like these types of questions are becoming rare, as they typically won't reflect how well a developer will do in the company. The closer the questions are related to the day-to-day activities, the better the company will be to work for. There is little point in asking about complex algorithms when all you will be doing is styling a front-end. Why ask about doubly linked lists when your chance of writing one is non-existent? These questions might have been applicable a decade ago.

The second final interview

In a small to medium-sized company, you might have one more interview. This is with the owner or upper management of the company. They perform one final check, ask a handful of questions, and check a couple of boxes.

I will stress this again; every company is different, and the larger the company, the longer the process. Google, Atlassian, and Facebook all have extremely difficult processes for landing a job. In fact, there are entire websites devoted to passing the technical interviews, but most people will work for small to medium-sized businesses with an interview process of equal length.

One final tip: if the interview process is awful, stressful, unorganised, rushed, or pressured, it will most likely not be a good company to work for. I recommend politely and respectfully telling them you are not interested. You may have little to no experience, but you should never compromise on your values, making yourself miserable in the process. A good company will pride itself on a smooth, rapid interview process.

A stress-free and organised interview process that is easy on the candidate and highlights that a company values their people is important.

How to Prepare

Once you've obtained an interview, it's time to prepare.

First, wear something professional that makes you stand out from the rest or gives you all the confidence in the world. I wore a suit at my first few interviews, which certainly helped. I don't wear them anymore, but times have changed with remote working. Thus, wear what makes you the most confident.

Second, attend with material to show off. If it's a sit-down interview in an office, you might have the opportunity to show off a cool side project you built (this is the ultimate show and tell). Show them your website if you have one or articles you have written (even small ones). Being more than just someone who sits down and answers a few questions will automatically put you ahead of the rest.

How to Ace the Code Test

In-person test

For me, this is the hardest part; for you, it might be simple. Either way, you need to be calm and methodical and demonstrate your thinking. The coding test isn't to see how good you are; it's to observe, at a minimum, that you can code and, more importantly, talk through your solution. Again, communication skills must be strong.

I won't sugar-coat coding tests; they are hard, not because the code is difficult but because there is a lot of pressure to prove that you know how to program, all while someone much more senior watches you type every letter. Some companies allow you to reference the internet, while others do not. Either way, you feel guilty forgetting anything, especially when you have to search for the most basic things because you have forgotten what a variable is. It happens; trust me. You shouldn't feel guilty; they will understand if they are a respectable company. You should be allowed to look up as often as you want because you will in your day job. Furthermore, you should be allowed to ask questions if things don't make sense.

In some interview processes, you pair program, which is by far the best method. Pair programming takes the stress out of the test because you guide each other. I have performed a test where I used the interviewer as my second pair of eyes since they can pick up on syntax mistakes or allow me to soundboard ideas off them.

When completing the test, the most important thing to remember is to talk through what you are doing, why you are doing it, and what you will do next. When I interview people, I don't care if the code works because the working code is not the goal. They need to explain what they are thinking and how they tackle the problem. People rarely finish the test; most people get around 80% through it, but that's okay because they get the job off the back of their communication skills.

Often, the tests are designed to be impossible to finish for 99% of people. You don't need to finish it; you just need to prove that you know what you are doing. Feeling stressed is part of the process, and it's easy to tell when a developer doesn't know what is happening, in contrast to one that is stressed. Do not be disheartened if you fail to finish because most are not designed to be completed.

Take-home test

The take-home test is an interesting one.

First, you must always question the expected time to complete it. I've been burnt before from spending vast amounts of time because I was under the impression I was almost complete. Once, I spent eight hours on a test for a company I was ecstatic to work at. I completed 75% of it and understood the technology, but I still didn't get the job. The problem is that some companies get you to perform free work. They give you a ticket they have and get you to complete it for free, then never hire you. So, clarify that you will not spend more than an hour on the test because you have many interviews to prepare for. If they don't respect that, don't work there; it's as simple as that. Typically, a take-home test will be much harder and require much more work, but if you time-box it to an hour, don't stress. Take a breath and complete as much as you can.

Your focus should be on clean, clear code, adding tests, and documenting what you have done via Readme.md or with your code via comments. Clean code and documentation rank much higher than completing the test. I can't tell you the number of times that people submit interview tests without adding code tests. The test itself will often suggest adding tests, so you must, but if it doesn't, you should add them all the same because that is part of the test. If you submit a test without code tests, you will not pass. You will pass if you follow clean code, add tests, add comments and complete as much of the solution as possible. It should not take more than an hour to complete. If it does, mention the time constraint and what you would have done if given more time. Inserting this type of information in the Readme.md is typical of a project in the wild, and this is the first place an interviewer will look.

The Most Important Part of an Interview Process

The most important part of the entire process is your communication skills. You can be a remarkable developer but won't work effectively in a team if you can't communicate. If you can't work in a team, you will be useless to the company and project. You will always win out against more experienced developers (to a point) by having fantastic communication skills, being happy and allowing them to see that you would be a stellar addition to the team. Teams will take the developer of lesser skill when it would compromise the relationships inside it by picking one of much greater prowess. I am not the best technical developer, but I can think, explain and communicate at a high level, which is how I acquire most of my jobs. I bombed in a few programming tests but still received an offer because I could answer all questions to an exceptional level. Furthermore, I communicate and show that I would be a great asset to the team. I have a low ego, don't take slights personally, and am curious about the stories of the people around me. Companies are rarely looking for an outstanding developer; they are looking for the best fit for a team. If you demonstrate that you are the best fit for a team, you will win the job over almost anyone else.

Being relaxed, calm, and comfortable highlights all your outstanding qualities. In today's age, we mostly do interviews over video or phone calls, which makes it somewhat easier. Performing interviews in your environment, space, or desk lowers your stress levels, allowing you to perform at your peak. Eliminating travel is one of the biggest steps in calming the mind and body. Breathe, know they would also be nervous in your shoes, and enjoy the process. I enjoy interviews because I talk about technology or recount stories of what I have done in the past. It's good to highlight interesting situations you've been a part of, your most successful project, or what you loved about a certain team. People love stories, and when they directly relate to the type of developer the company could potentially have in their workforce, you rise on their candidate list.

When interviewing candidates, I pick those who I would most like to work with over the best candidate. I have a minimum level they must hit. This gets raised or lowered based on how passionate they are, which directly correlates to how much time they spend learning and improving. Then I look at their communication skills and how well they can discuss technology and the interpersonal situations that arise. It isn't about the answers to the questions; about teasing out the personality and the thought process. If they can convince me to change or question my ideas on a subject, that's even better. If someone has poor communication skills, I will almost always not pass them because it doesn't matter how great their development skills are. Furthermore, if they can't work in a team, they cause more harm than their output recovers. I'll say it for the 100th time; communication is key and will get you further than any technical skills.

Some of My Worst Interview Experiences

I was asked to perform a double test in one hour in two languages (one front-end, one back-end) in a Word document. I wasn't allowed to look up anything, I didn't have an editor, and I didn't even have formatting. The test description was confusing, and I was interviewed by someone with such a thick accent that I had to ask them multiple times to explain what they were saying, eventually giving up and guessing. I did not get the job, despite desperately wanting it. I was extremely disappointed, but at the end of the day, it was probably not a good place to work, and I found something for more money and a better culture anyway.

Once, I attempted the test live and just forgot everything. I couldn't remember how to assign a variable or what a loop was, but this happens sometimes. The live tests are hard not because the actual code is difficult but because it's a new codebase you have never seen, and it might be using technology you lack experience with. Furthermore, it's just stressful! Don't be disheartened if you perform like a person using a keyboard for the first time; I've done it numerous times. You just have to practice, which means applying for more jobs and going through the process more times.

I interviewed for a big company in Australia, and one of the steps was a timed test, which was stressful. My friend had been through the process and said the test was simple. Despite not knowing the technology involved, he chose the front-end test and scored 95% (it was automatically scored) and even finished early. I decided on the back-end test, which I'm more comfortable with. I scored something in the range of 50%, with a pass rate of 75%. Furthermore, I didn't even come close to finishing the test, and I knew what I was doing. It was simply so long that no matter how fast I went, I would never have finished it. I described the test to my friend, who was shocked at how different it was from his. I asked for a retest as I found this quite unfair, but I never attempted it because I got another job with a company with a much better interview process and ultimately became a great workplace. As my wife always says, 'Everything happens for a reason.'

I interviewed as a senior developer for a company offering me a huge pay increase (50%) to what I was currently on. I had made it through the first two interviews, completed the test, and everything was smooth sailing. The final interview was with a few people, namely the tech lead, who was my future boss. This tech lead asked one question in an hour and a half and said perhaps 20 words. This is not what you want from your tech lead and the advocate of your team. The next day, I received a call from the recruiter handling the process, and he said I had passed, and they gave me the offer. Not only was it under the value they had advertised, but the reason it was under that value was also that I would be the highest-paid person in the company. When someone says that, run immediately; it means you will never get more money, and the culture is awful. It's a tactic often used to lowball people into taking less money, and it's a slimy play. Any company that promotes a certain pay and then lowballs you for a reason other than thinking you are almost there and wanting you, so offering slightly less is likely to take advantage of you.

The final straw was when they attempted to pressure me into signing a contract on the spot, which was a huge red flag. I hadn't even read the contract, and they wanted me to agree to it verbally and sign it immediately. Never sign a contract without reading it cover to cover multiple times. Always get someone else to read it as well. Some dodgy contracts out there can be downright illegal, although good luck trying to prove that. One of the clauses in this company's contract, roughly translated, was that they could monitor all activity on your laptop at all times. Now, all companies do some surveillance. However, I had been burnt in the past with companies that install software on your computer that records everything you do, takes screenshots, and logs all the keys you press, and I was not having it. Being constantly monitored like that is suffocating; you always fear that you might say something or do something your boss can use against you. Never work for a company that carries itself this way; they are never a great place to work.

Some of the Best Interview Experiences

In one job I obtained, the entire interview process consisted of an hour-long phone call with the tech lead and a sit down with the tech lead and a developer. We discussed various technologies and situations for two hours during the sit down. It was easily the best interview process because all we did was talk and discuss the technology they had, what I had done, and how I would solve problems. Not a single line of code was written. It was relaxed and stress-free, just the way it should be.

Typically, the more streamlined the process, the better the company. In my experience, the interview process strongly translates to what you experience on the job. Don't take the job if you have a bad feeling about the company.

Another great interview I had was with a company that had a three-stage process:

- **Call with an internal recruiter (30 mins)**
- **In-person test and Q&A (1 hour 30 mins)**
- **Q&A with the owner of the company and upper management (60 mins)**

This process was streamlined to a point where the time from when I heard about the company to when I got the job offer was about a week and a half. I have never been through a process this fast and fun. The first call with the recruiter was easy and relaxed, and it was about meeting my expectations and theirs and how I communicated. The test and Q&A were a bit more stressful because it was a live test, but it was as relaxed as it could be. They sent the codebase in advance so I could become familiar with it. They let me use my editor on my laptop, and the questions were clear with no right answer, and they allowed me to talk through my thought process and why I thought the way I did. The final interview was relaxed, even with the presence of the CEO. They highlighted the company's values, selling me on why I should join. They let me ask questions during all three stages of the process.

A great interview process will attempt to sell you on joining because it's just as much a sales pitch for you to join as it is for them to hire you. The interview process is a two-way process that allows people to sell themselves and, ideally, the company to sell themselves. At the end of the day, they attempt to attract the best talent, and with such a highly competitive industry, this can be difficult. This is fantastic for you; it forces the interview process to be streamlined and to cater to the candidate.

Interviewing can be stressful, difficult, and insulting. There are great and horrible companies; you can tell what kind of company it is from how they treat you in the interviews. You must seek the best companies in how they treat you, not how they look on your resume. Continue to apply for jobs, and you will get an interview eventually. When you do, relax, breathe and don't take the first opportunity that comes knocking if it doesn't feel right.

CAREER PATH

What Is the Software Ladder?

There are many paths one can take to climb the engineering ladder, and it's perplexing since most companies change the names or have different alignments between companies.

Typically, it will follow this pattern:

Graduate: part-time position for someone still in university or just graduated, but the company wants to avoid paying them a good salary.

Junior: the entry-level full-time position, where you typically start and stay for a year or two, depending on how well you perform and how the company progresses their employees. You often require help initially, but you start to action tickets on your own and tackle bigger problems.

Mid-level: someone with a couple of years of experience and who tackles larger problems without much help.

Senior: someone with three-plus years, depending on their skill level. They can handle the tough tasks in a team.

Lead: someone with five-plus years. As a lead, you must run a team as a tech lead or manage multiple teams. Typically, you have a team of 3-10 developers, perhaps two teams of five. You won't be coding much at this point; you will mostly be directing people, helping people do things and getting unstuck. You may occasionally have to tackle a difficulty issue, but often the only code you work on directly will be tasks that won't block anyone. Furthermore, you attend countless meetings, which can be the bane of your existence or your calling.

Principal: someone with 15+ years of experience. You seldom program. Typically, you are the head of many teams and have multiple leads directly report to you. You interact a lot with clients if you work at a consultancy or direct large projects from a level above a lead, thus managing multiple leads and teams.

Engineering manager: someone with 15+ years. This can often be adjacent to the principal level, but instead of being on the tools or higher levels in projects, you manage many developers, track how people are travelling, and manage career progression pathways and reviews. An engineering manager never programs; their days are jam-packed with meetings instead.

No two companies have the same grading structures; some skip lead and only have senior to principal. Other companies might have alternate names for their levels. At the end of the day, the paths are the same. You can stick to a strict engineering path or branch out into management roles. It all depends on your strengths and weaknesses. You might get to lead a team as the head technical developer but find it mediocre and just want to program, and that's great. Don't go further down management; you can continue to become a senior engineer, earn lots of money and work on great projects. Just because you stop at senior doesn't stop your growth or your income potential. It just means that you will be the point of call in a software team, asked to solve the toughest problems and fix the most obscure bugs. If these are your strengths, you will perform extraordinarily. As the senior, you're required to help others in the team, but you will struggle if you are not good at this. However, as the senior, you still spend most of your day programming. If you want to split your time between programming, teaching, mentoring, leading and delivering a project, you should aim to become a lead or higher. As a lead, you become the point person on the team from the technical side. Typically, you spend your day performing 20% programming and the remainder in meetings, mentoring and helping the developers with issues.

How Long Does It Take?

It can take as long or as short as you want it to. If you put in significant effort and pick things up fast, you could be at the lead level in five years. If you just bug-fix software and put little effort into your work, you might make senior in 10 years. To go beyond senior, you must show you can lead, mentor and run a software team. You often don't become a lead just because you have been a developer for a long time. It's the first step in a management role, and it requires significant effort outside your programming to perform successfully. The further you get up in the managerial roles, the less programming you do, the more meetings you have and the more mentoring you will complete. If soft skills and getting projects delivered are what gets you going, then these roles are for you. In many smaller companies, a tech lead will also have the power to decide on the entire technology stack, how it's architected and the team's patterns. In these smaller teams, the most senior engineer gets to make these decisions. In bigger companies, you might need to become a solutions architect if you want to design a system. You will probably never build it, but at least you figure out how all the pieces work at a high level, and it's now up to software teams to build it. Occasionally, you get dragged into a meeting to discuss your architecture or to make changes, but besides that, you won't have the chance to build it.

These levels all take different amounts of time and blur between companies. You might find that you are a senior engineer at one company, and then at another, you're the lead but design the system as well. It's all highly company dependent, and usually, the smaller the company, the more the lines blur, and the more you have to wear multiple hats. If you're up for the thrill, working at a small company gives you exposure to each level, which provides you with a glimpse at the various levels.

Contract vs Permanent

Permanent

Most people will be permanent employees, meaning they get a salary and benefits for working a 40-hour week. This is the simplest form of a job. You're entitled to paid leave and potentially other benefits. This is fantastic when you begin your career or just don't want to worry about finding a new job every 6-12 months. You can stay there, get mentored and potentially be promoted. I have often found that you are forced to go above and beyond, and because you are a permanent employee, you don't get paid extra for working overtime. If you find a great company, this won't be a problem. If you find a poor company, this is awful. I have had companies squeeze as many waking hours out of me as they possibly could.

Contract

Being a contractor will pay about 2-4 times more per hour, but you won't get any leave or benefits, and you have to supply your insurance and laptop. You can be fired without cause with only a day or two notice, and a contract will last between 3-12 months. With great upside comes great downside. You can earn vast amounts of money in a short time, but you can also spend months without work if you can't find any other contract position. On the plus side, you can take extra time off each year by not picking up another contract until you want to. This means you could work for six months, earn what you do in 12 months, and then take those other six months off. You could do this every year if you wanted to. My only issue with this is that you will not progress in your career as fast, so I would only recommend swapping to contract work when you feel you have peaked or are no longer burdened by the pressure to progress up the corporate ladder.

Contract work can give you great freedom if you accept the drawbacks, are good with your money, and understand that you must put money aside for tax each year. Moreover, as a contractor, no one is doing your taxes for you. You need to do this yourself. This means you must put money aside from each paycheck for your tax return. Depending on your country and income level, this could be as little as 10% or as great as 60%. You must ensure you understand roughly how much you need and keep aside to avoid being caught. For example, if you earn $150k as a contractor over 12 months and must pay 20% for tax, you would have to keep aside $30,000, which is fine if you do so each week. However, you will be in for one sore wallet if you get to the end of the year and haven't kept track.

As a contractor, there are just more things to keep in mind, but it's not hard if you get an accountant to run you through it and set everything up to be automatic.

The last great thing to remember about being a contractor is that you find more things you can claim on your tax than an employee. Again, it's highly dependent on your country and tax bracket; however, you have many more options for saving on tax as a contractor you don't have access to as a permanent employee. Plenty of items become business expenses, which allows you to save a great deal of money on tax.

Enterprise vs Start-up

One thing I don't see talked about enough is the differences between a start-up and an enterprise. I have worked for both, and they are very different. Ultimately, the choice between the two depends on your personality type. I'll preface this by saying that I am not in favour of an enterprise. I think they have plenty of benefits, but for my personality, I find them boring, full of red tape and quite often incompetent. If the glacial pace of a company is something that hurts your soul, as it does mine, an enterprise is not for you. Conversely, if a start-up's rapid, heart-racing, stressful nature is too much, stick to an enterprise.

Enterprise

These are your typical gigantic companies that employ thousands of employees. They don't have to be Google or Amazon, but they can be big banks, insurance companies, etc. There is a variance between a technological enterprise and a regular enterprise. Technical enterprise companies are superior for developers as they build technology. They still suffer from many of the pitfalls of big business, but they simply solve more technical problems. Non-technical enterprise companies are places that make or produce something apart from technology. They have intelligent people littered around, but they are not focused on tech, and these places create more frustrations for developers. Technology-focused companies have the drawbacks of a large company, but at the very least, they can offset these by understanding how to be a technical company. Non-technical companies have no concept of technology; it is an afterthought. As a developer, this could dishearten you to no end.

Pros

You can learn incredible things from these places as they have solved most issues, have brilliant people who have been around for a century or more, and have many opportunities to move around. You might start fixing bugs in a small team for some obscure application, then move to a team building an entirely new product.

At an enterprise, you have room to grow, the career ladder is planned, and it's sufficiently understood how to progress. Each level has markers to reach, so if you fail to reach them, you fail to be promoted. However, if you fail to meet the markers, they have support on how to reach them. You can always find someone who knows someone who knows something, which means less time being stuck and more time learning and progressing.

If you don't enjoy the work or the team, you can request a change of team and project without leaving the company. There may be opportunities for internal promotions in various departments. This allows you to move around the career tracks, and you can even leave engineering entirely.

Your days are much like walking into a museum; one day, it's ancient codebases, and the next, it's modern, which allows you to obtain a broad understanding of a language. Observing what Java looked like 10 years ago compared to today can alter your perception of a language, and your understanding deepens along with it. Your skills become highly transferable, as almost all big companies use the same technologies, at least in some parts of the company.

Best practices are followed more stringently, standards are set, and you're required to deep dive into testing. Everything is set up for you, which means as a first-timer, you won't have to go and set up a complex testing framework from scratch or construct complex parts of a system. Instead, these things were standardised years ago by people who have long departed the company.

Cons

Every company has cons. However, you should be aware of commonalities with most large companies. Large companies can be dull. For instance, there are teams dedicated to fixing bugs all day, every day. If you find yourself in one of these teams, which can regularly happen, given that they tend to put junior developers on these teams, you will be bored if bug-fixing isn't your thing. Most problems are solved, and there are clear and fixed guidelines for everything. For you, this means that you can't experiment, you won't be able to use a new and exciting bit of technology, and you might be restricted to what the company already uses.

There is a metric tonne of red tape; everything has to be approved within an inch of its life, and everything requires 800 pairs of eyes. You can't make a decision without another's approval, namely your manager's manager. The flip side to this is following every rule set out, but a governing body goes and changes the rules. I have seen it in the past when someone high up decides that the technology you have been using isn't to their standards. Then they force your team to conform to the new standards, no matter your project's phase. I have seen this happen before, and it ends badly.

With larger companies, there is also a greater chance of useless managers who don't provide any value. They get into these positions, stay there and make your life hell; it's just the nature of large companies. Awful workers slip through the cracks, and as a result, they sometimes get into management positions. When they do, this is when it causes all kinds of pain. They dictate what you can and can't do and change the direction of teams to suit their needs. Everything revolves around them and how they can progress. Unfortunately, these situations, although rare, can happen.

Start-ups

Pros

Start-ups are crazy places. One week, you could be user testing; the next, you could deploy your code, neither of which you have done before. This is the world of start-ups; you're asked to wear many hats and perform well beyond the call of duty. You learn an insane amount, and it's frequently trial by fire, which means, here is a ticket, sort it out, and talk to as many people as possible if you require help. Typically, it's not the lack of people that want to help; it's just the lack of people, which means you have to get cracking. I have had a few instances where the ticket was much harder than I should have attempted. However, I stuck to it and got there in the end whilst learning and stretching my skills. That's the thing about start-ups, you have to stretch far beyond what you are comfortable with, which is great if you can handle it.

In start-up land, you usually get free-range to make all the decisions and try out fascinating and bleeding-edge technologies. Rarely does management intervene in the technical decisions, and seldom is there red tape. If you want to push your code to production, you just push it.

Cons

Start-ups are brutal. There are no off days, you never get to relax, and you often push yourself to the limit. Someone who doesn't find this appealing should not aim to work at a start-up. You will be asked to work more than 40 hours and complete work that drives you crazy with its complexity. Nothing in a start-up is a solved problem; everything is new ground. Often, there isn't someone in the team with experience in this matter, so you must figure it out yourself. You won't have clear management structures; you won't have a career path. Performance reviews are an afterthought, so you will have to fight to prove why you deserve a promotion, which can be easy if you push a product to market. Typically, you have to arrange the performance review, not the other way around.

Consultancy or Agency vs Product

Consultancy/Agency

I use the term consultancy and agency interchangeably because, for the most part, they are the same thing. There are some slight differences, but most companies blur the lines between them, making it moot. However, an agency is usually a team of developers who are hired to produce a product. A consultancy is where a person goes into another company to either work in the team or offer advice and suggestions. In my time, they typically work the same, meaning you build something for a client.

Working at a consultancy is all about client relations since you spend a lot of time building a relationship with the client to obtain future work. You must perform to the client's standards or better. You get little say in the project's direction, what features are created and how things are done. Your goal is to build a great relationship, drive as much change as possible, and keep the client happy. You are often put into one of their teams as a developer or an advisor. As an outsider, you have the power to make observations about how things are run, and you can provide suggestions. Working with clients is rewarding and frustrating; they want whatever they want, which often makes little sense. You go into the gig expecting to make a lot of change, but at the end of the day, you make almost none.

Pros

Working in an agency can be spectacular. You get to build new things from the very beginning. You typically don't get to select the technology or decide what features should or should not be there, but you get to build it. You get your hands dirty solving problems and maybe even get to choose a few things; it really depends on the client. I have seen some agencies pitch the entire project, including the technologies and features. Often, this is how work is won. A company puts out a piece of work, and people bid on it with proposals on how it should be done and with what technologies. As a developer, you won't be a part of much of this, your boss's boss will make the calls, and your team will execute them.

Consultancies are where your communication skills are useful; you can't be a good consultant if you can't keep the client happy. On the other hand, you can't be a good consultant if you can't program. This makes it doubly important to be good at programming and communication. These challenges propel your career; you can get far on great communication skills alone. In fact, I would say I have won more jobs with my level of communication than with my development skills. You can be the best developer in the world; however, if no one knows, you won't get very far.

In a consultancy, on average, you can move clients every 12 months, which is great if you get bored or win over a client easily. Furthermore, you get a diversity of thought, work and clients. If you need a break from a client or a type of work, you won't have to leave the company; just leave the client. The tough nature of client-based work will always be hard, but with hard work comes great satisfaction.

Cons

Often it can seem that you get put into a client as a consultant for your software expertise, yet regardless of what you say, they still make all the wrong calls. Often, you can't change their minds, no matter how hard you argue. This can be frustrating and lead to burnout and disappointment with your work. It's hard work being a consultant because consultants are the first people out the door when something goes wrong. The worst part is that consultancies often rely heavily on a few clients. If a client isn't happy anymore, they remove the consultancy, and you end up sitting on the bench, twiddling your thumbs until the company can find another place to put you. This can be a great time to learn and recoup a bit. However, the more days you aren't billable for your company, the worse your performance review, even if they don't directly blame you, which they shouldn't. There is a lot more pressure in an agency to deliver a project; you get the blame when it doesn't go right and rarely get the credit when it does. This can wear you down. The deadline pressures are immense. Diamonds are created under enormous pressure from carbon atoms, like a consultancy, and you are under the same level of pressure. Deadlines are not moved for any reason; your company is paid treasure troves of money to hit them. The team might raise dozens of reasons for the delay, but the company won't care; they just want you to get it done. This involves many overtime hours and a lot of stress for the developers.

Product

Working on a product means your career is in your hands, and you can take it in any direction you want. You can focus purely on the development side and head to management. You might even get out of software engineering altogether, but one thing that never changes is that your career is in your hands, and you can do what you wish. Furthermore, you must build up the career capital to get where you want to go. If you want to spend less time working for someone else, you need to get good enough as a developer to earn that right to work fewer days or hours. Better yet, branch out on your own and choose the number of hours you wish to work, devoting the rest to your endeavours. The more you focus on the direction you intend to go, the better your career becomes and the happier you become. Freedom is your ultimate happiness, for you can pursue anything you desire, making way for what matters most in your life.

Pros

Working on a product means you get a slot into a team, you get to know a certain section of the product in-depth, or you might build a new product. By the end of your tenure, you will know just about everything there is to know about the product and business, and you will become an expert in the language and tooling. You will be asked to go deep into the company's technologies, and no matter the knowledge, you know there will be someone who knows even more.

If you join a small to medium-sized company, you will hopefully watch it grow from a small idea to a bigger product in the market. If you work for a larger company, they will most likely have a large customer base, which means your code could be used by millions of people worldwide, which is a feeling few other fields have. You will learn a lot about an industry. For example, you could be working for a gambling company and might love gambling, so this is the perfect fit. You can apply your gambling knowledge to your everyday work, surrounded by other people with the same interests. Maybe you choose a company in a field such as finance to learn more about the field, and after a year or two, you will know a lot more about finance and how it works from the inside out. Looking behind the curtains to see how it works from the inside can lead to further opportunities. This can be great if you want to pivot your career to something else or build something yourself based on your knowledge at one of the companies.

Cons

The biggest downside to a product is diversity. Sure, you can move teams, but for the most part, you build something directly related to that core product. If it's banking, you build additions to their mobile app or bug-fix their back-end systems. You can find yourself stuck in a bug-fixing team, or if you work at a company with regulatory requirements, you can get stuck ensuring all the code is up to scratch each year. That might be your jam, but it will probably not be that exciting for a first job.

If you don't like the product, you must find a new job. You can't request a different client if you don't like the boss/client. If you work at a company that builds software, think Google, then you're probably going to have more interesting work, but if you work at a bank, there is a lot more boring, stable, bug-fixing work over new and innovative projects.

It can also be difficult to stand out in a product company if it's larger because you get placed into teams and potentially forgotten about, so you must push to stand out. If the product company is a small start-up, your entire work could be for nothing if the company folds. You might spend two years building a new application only to find out that the company shuts down unexpectedly because it ran out of money. Again, if you work at a big company, they might dump the entire project mid-way or after it's released. Budgets or the board of directors could change, and then the priorities change and your team is disbanded. As a result, you're sent to another team, but you still feel the sting that the product you spent months of your life on is simply no more.

It's All Personal

No matter the path you take, it is entirely personal. The company you choose, the time you stay at places, the roles you take and the level of hardship you are willing to put up with. It will all determine how you feel about each avenue. The biggest factor is that all this is highly dependent on your personality. Some people suit the fast-paced lifestyle of start-ups, the possibility to make it huge and get a stack of money in return. Others prefer the consistent workload of an enterprise with lots of support around them. They never have to work too hard but are never out of work. You can coast in an enterprise or push yourself to climb the ladder. There are guide rails everywhere, so you don't get to make many decisions. On the other hand, if you like to get deep into things, figure it out and push yourself beyond your limits, then a start-up might be for you.

Agency life is also an interesting avenue, or a product company might be better suited. Do you like to stick to one product and know it and the company inside and out? Do you like to jump between projects, learn a diverse set of skills, picking up something new with each client? It's all entirely up to you and what you prefer, and you won't truly know what you like until you try.

My only advice would be first to try the path you believe matches your personality best; secondly, stay at least a year; and lastly, reassess after that year. If the positives outweigh the negatives, you know it's for you. If you simply hated it, try something new, swap from an enterprise to a start-up, give it another year, and reassess. Gone are the days when you're required to stay put at a company. I move every 12 months, and it has not affected me once. Move around and explore the various companies and styles on offer. No two companies are ever the same; as they say, 'Variety is the spice of life.'

TAKING THE ADVICE OF MORE EXPERIENCED DEVELOPERS

One thing I used to struggle with, and to an extent I still do, is taking the advice of people with more experience than me. In life, especially as a software developer, you come across people with more experience than you. No matter what the area, there is someone who has been there longer, worked harder and failed more.

When I began my first job as a developer, I believed I could learn 100 times more from the seniors than from any textbook. As a result, I would seek their thoughts on various topics. I would ask a thousand questions just to probe their minds. I don't think I'll ever forget this, but maybe three months into my first job, I created a PR for something I was working on. Now it was a bit of a big PR, as some requirements had changed along the way. It had gone from one ticket to three. Sure enough, it became a bit of a mess, with bits of one ticket here and bits of another there. No one was quite sure what code was solving what problem. This, of course, generated many discussions. One hundred comments in, and I can say I wasn't exactly full of confidence. My PR had been torn apart by three developers with more experience than me. After feeling down about it, the three reviewers told me it was not personal but about improving. At first, as you can imagine, it didn't exactly feel that way. However, I look back on it fondly as it made me a better developer. Gone are the days when I created PRs for three different tickets combined into one monolithic beast. Instead, it's one PR for one ticket, and sometimes tickets need to be broken down into smaller parts, each with its PR.

I could have cried and gone running to my mother, as many people can attest to, I'm sure. Getting your code torn apart piece by piece isn't what I call confidence building. Yet, when we face adversity, we become better. These moments build us up. My failure of a PR made me a better programmer. If it didn't happen back then, it would have happened years later, which would have been worse since rookie mistakes should not be made after so much experience, so get them over with early.

Non-technical Interactions

When working as a consultant, non-technical interactions become a regular occurrence. As a junior, I never had to do this.

Interacting with non-technical people takes time to develop, but you can accelerate the process by listening to the people around you. If you're in an enterprise for a product company, you might find that everyone is technical. You might also find yourself with a non-technical product manager (someone who oversees what features should be built and represent the business). You must master the skill of translating your technical knowledge into layman's terms because it's invaluable. Take, for instance, my boss in my first year, who has been working in this field for 20 years. He knows a thing or two about interacting with non-technical people, the dos and don'ts when discussing design decisions, and how to explain why something should be done a certain way.

The opposite of that is time estimation, and this one is tricky. You will never master it; just slowly get better until you're correct. As you develop this skill, you better understand the common pitfalls that could drag a ticket out. As anyone who's dealt with a client directly, they know that if they don't have any programming experience, it can be a nightmare to explain why adding this simple function could, in fact, take weeks and why it would mean a complete rewrite of a major part of the system.

These situations can become difficult. It will take many years and many client interactions before you're comfortable. However, you can speed this up tremendously by soaking up the surrounding knowledge. Your lead must likely deal heavily with non-technical people, so take their knowledge. Spend some time discussing their tips and tricks and any major pitfalls you could avoid. Lunch might be the best time to talk, so probe them about their experiences, and soon you will hear the good and bad. Tell them about a situation you are having or ask them what area you could improve on. They might offer advice or tell you about a similar situation they had and how they dealt with it. This is how you get insider tips on what it's like in the industry. You won't believe what some people have experienced in their careers or their paths. I have worked with people who went to university, never finished university, did boot camps, changed careers, were mothers and fathers, worked part-time, full-time and have come from all sorts of countries.

Improvements

You need to take every opportunity to improve. The beauty of working in a team is that there are plenty of people to help you achieve this. The avenues of improvement are via PRs and pair programming. These are opportunities to grow, so take full advantage of them. The more people who see your code, the more you can improve. Obviously, not everyone is willing to help, but for the most part, teammates will review your code and point out improvements. Someone simply reading your code can catch a stray piece or some leftover comments. You often find that people have suggestions on how to write a piece of code, and every so often, they know a special trick in the language to make things easier; these are the best kinds of reviews.

You should also be reviewing other people's code. The more you review their code, the more your code can improve. In reading the code of others, you can see what reads well and what doesn't. This is typically difficult to observe when you write it, as you are too close to the problem. When you read someone else's code for the first time, you have little context about the problem, making you the best person to pick up readability and understandability issues. It makes little sense to keep code that only the owner can read. Instead, you provide small suggestions for improving it as someone new to the code. Inevitably, you or someone on your team is required to maintain this, so it must be easy to do so. You should always remember that your code will be read many more times than it is written, so you must make it easy to read. As you continue reading more code, especially that of more senior developers, you will notice good and poor code quality. It becomes imprinted in your mind, subconsciously making you better.

You can often tell when someone has spent a lot of time programming by themselves as they tend to build bad habits or build in a way only they can understand. This is natural because they are on the problem. Thus, it's easy for them to understand how it works. However, when someone else attempts to read it, issues arise. It can be impenetrable to read others' codes, which is why it's critical for others to read them early on in the process. Think back to writing an essay in school; first, you write a draft, the teacher reads and marks it, and you fix the issues. This is what writing code and getting a PR reviewed entails. Code is never complete the first time you write it, much the same as an essay; it requires continuous revisions. You build the skill of showing others so that it makes sense to someone besides yourself.

Showing your code to others, especially those more experienced than yourself, is how you improve. You will mess it up a thousand times, but as long as you keep trying and getting eyeballs on your code, you will keep improving. When a renowned writer writes a book, they don't smash out 100,000 words, send it to the publisher, and it's printed. No, it goes through many stages of editing. In most cases, the editing process takes longer than the writing process. In fact, it is speculated that to release a book of a high standard, you endeavour to spend three to four hours editing for each hour of writing. This can be directly applied to your code; you should spend more time editing and improving it and listening to feedback than writing it. The more feedback you get, the better you get. The more uncomfortable you feel, the more you will improve. The root of stagnation is complacency.

TAKING THE ADVICE OF LESS EXPERIENCED DEVELOPERS

This is about the more experienced developers in your life and how you grow the most from them. However, you can learn a lot from anyone. I have been gobsmacked talking to someone with only six months of experience, blowing me away with some fragment of code or language feature I had no idea about. Everyone has something to teach; you just have to listen.

If you ever find yourself arrogant enough to ignore those around you, you are in more trouble than you realise. You can learn something from everyone, even if it's not development related. People come from diverse and interesting backgrounds, so you just have to listen; they will tell you a lot. I have learnt many life lessons from talking to developers around me. There was one woman I worked with who was particularly smart. I found out she grew up in Russia in a one-bedroom apartment with no toilet for a family of six. It was an extremely rough upbringing, but she worked extraordinarily hard and managed to move to Australia, attend a good university, and get a high-paying job as a software engineer. You wouldn't have suspected this by looking at her. You wouldn't discover this without talking to people and asking about their stories. Listening to people's stories puts your life into perspective and gets your head out of the code.

When you sit there stuck on a chunk of code for days, getting frustrated and angry, you can think about how you have a well-paying job, somewhere to live and food on the table. Gratitude helps you put things into perspective and allows your mind to relax, and then you solve the problem, or maybe you don't, but who cares? You get there eventually.

WORK-LIFE BALANCE

Never Work More than Eight Hours for Your Employer

I often see junior developers taking a tremendous amount of work home because they think that since they got stuck on something, they are failing to deliver in the exact three days and five hours they estimated and that somehow it appears they don't know what they are doing.

I also felt this way, and I would take work home many times. But do not do this; it's something you must learn. I still struggle with this today.

It's admirable to take pride in your work and want to prove to everyone and yourself that you can do the job well and that you deserve that raise and promotion. However, it's rarely rewarded. In fact, the more you do it, the less it is rewarded, and then you just end up burnt out. Trust me, do not work more than eight hours. Put in a good day's work but leave it at that. Come rested and ready to go tomorrow. It's not a sprint; it's a marathon. Take your leave, take sick days, and don't work overtime unless you get paid for it and have the choice!

I have worked a lot of overtime, and it's seldom rewarded. Typically, it results in hating the company and leaving not long after. A sign of a bad company is one that makes you work overtime to get the job done; it's a systemic issue that's hardly representative of a good company.

There Are Companies Out There That Won't Force You to Work These Hours, so Work at Those Companies

There are companies out there that won't make you work these insane hours; some mandate that you never work more than 40 hours a week, which is a standard working week.

Leave if you're at a company that doesn't treat you how you should be treated. Do not put up with toxic companies, wasting your time when there are plenty of jobs for great, hardworking individuals. Know your worth and know that with a solid resume, a website and fundamental skills, you can find a well-paying job that won't work you into an early grave.

Typically, at larger enterprise companies, you don't work more than 40 hours, the benefits are great, and most people are there to cash in on big checks and reasonable hours. If this type of work is not to your taste, like me, simply look for a smaller company that treats you well. They exist; you just have to find them. If you are looking for a fast-paced start-up lifestyle, you must accept that many pull insane hours. This is great for some, not others, so choose the pace that works for you.

If You Want to Work More, Work on Your Projects or Self-learning

You're young and full of energy, so why not put in a few extra hours for yourself? Work on side projects, learn something new, learn a new language, or read about something that interests you. Don't spend it at a company that won't pay you extra or credit. If you spend one additional hour learning something new rather than working on something at work, you will find that it makes you much better at your job anyway, which will have a much greater impact on promotions and raises than working that extra hour.

I can't tell you the number of times that something I learnt in my spare time has surfaced in some way for my job. It might not be next week; it could be years from now, but it eventually appears and comes out of nowhere too. One minute you are struggling with a problem, and then it smacks you in the face. You have dealt with something similar before. There are many times someone has come to me and said we need something to solve a problem, and I could give them a few options, weighing the pros and cons until I found a solution. Anything from a front-end component library tool to a testing framework to hosting platforms. It can be anything, but it's about having that knowledge bank. You can do this by simply reading about and programming anything that you fancy. If you're no longer interested, stop and find something else intriguing. If you're too tired from work or can't think of anything, don't worry; just leave it. There comes a time when you have a problem to solve, so search for it, understand it, and build a toy application to solve it. This is how you get a wide understanding of the field. Never force yourself to do anything; you won't retain it. Burnout will knock at your door when you push for too long.

Even if you spend time on projects for yourself, doing too much of anything is not a good thing, and you can still get burnt out no matter how much fun you have. Don't listen to these hustle culture speakers telling you to work 100 hours a week and never take a day off. They push the notion that you can't possibly be burnt out if you love your work. However, this is not true. Regardless of what you do, if you do too much of it, you will become burnt out; it's just a matter of when.

Pick up a hobby that isn't on a computer; it's important. I picked up a new language once, not a programming language, but French. I am not proficient; it's taken a long time, and I'm only semi-competent. However, it was one of the best things I ever did. Not only that, but I saw things in a whole new way. From a software perspective, it made me think about language conversions between applications and how some apps didn't translate anything while others did. Seeing these applications I used every day from a different perspective was fascinating.

It made me think more about how you work with languages in your software. English is a massive language and the most common worldwide, but it still doesn't have the most speakers; it's not even in the top two. Many people don't speak English well, can't read it, or choose to use their native language. Some older people simply don't have the energy to become proficient in English but still enjoy technology. These are all things I didn't think about as a mono-language person.

Picking up any hobby that isn't in front of a computer is great for your health; it takes you away from the code and the screens and puts you in front of an entirely different challenge. It creates new neuron pathways, relating information you have learnt from other endeavours to this new one. The most challenging part is listening to what truly interests you. You might suck at whatever you choose to do for ages, but once you do it for some time, you develop a level of competency, and it will expand you as an individual.

When you reflect on your life, you don't want to regret not learning to skydive, play the piano, etc. Life isn't just about work, so don't waste your time working every minute of the day.

Building Career Capital

One of my favourite books of all time is *So Good They Can't Ignore You* by Cal Newport. He proposes the idea of building career capital to give you the freedom to pursue whatever you want. I find this to be such an important point. The industry we find ourselves in is highly paid and with more jobs than people. With this in mind, it's a great idea to build that career capital, experience, resume and social network that allows you to pursue other avenues if you so desire.

Career capital is about finding the routes you wish to take and leveraging experience and network to make it happen. You might find that after 10 years, you wish to work three days a week. With 10 years of experience, you could leverage your expertise as a contractor and get paid well but have four days off each week. This gives you the freedom to pursue whatever it is you want to pursue. On the other hand, you might want to create a start-up with an idea you had. Not only do you have the skills to do it, but you would have the network to find people who are interested in helping. Your career capital gives you the safety net of experience and network to achieve these goals.

WORKING WITH
GREAT PEOPLE AND
IN GREAT TEAMS

Finding great people is one of the hardest things to do at any company. You can hire people easily, but hiring great people is hard. Some companies do a fantastic job of this; they only pick people that complement the company, even if it means passing on someone who is technically better. These companies realise the importance of working with great people. People are the lifeblood of a company, so if you mess with this, you mess with the company's success.

Bringing in great people is difficult; they want the top salaries, they have more experience than you need for the role, and there is fierce competition. You must have a great recruitment team to do this, but when you do, it pays off in leaps and bounds. I have worked in companies with good and bad cultures. The difference between the two is like night and day. Working with people who don't wish to be there is draining and, as a junior, soul-sucking. You want nothing more than to work, learn and improve, but you have people around who don't want to be there. They're sad, miserable, and checked out, which rubs off on you. If you stay in these companies too long, you will feel the same.

You become a product of your environment, which is why the best companies spend so much time, money, and energy creating great office environments filled with great people. I worked for a large payments processing company with roughly 50 engineers for a stint. Only one person wanted to be there, and the rest were cashing in on the enormous contract rates. The turnover at the company was ludicrous, with most people only lasting around 10 months. This was a huge inconvenience when you tried to get help on a legacy system; not a single person had worked on parts of the codebase because most people had left. The office was depressing, looking like the cubicle hell of the 90s. One of the most striking examples of how terrible it was, was when the financial report came out. They had a meeting that everyone in the company attended. They discussed how fantastic the company was performing. At the end, it was highlighted that the company made a 33% profit margin, which resulted in millions of dollars, yet the office looked like it was two weeks away from falling apart. It wasn't until later that I discovered that managers even struggled to get approval to buy pizzas for their team! How can a company not spring for some pizza when they make millions in profit? There was a reason great people didn't work there; they wanted to work at great companies. Thus, any company that aspires to be great will hire great people.

What Makes a Great Person?

It's a pretty vague statement to say 'hire great people', but when it comes down to it, it's about hiring people that complement the culture, work well with others and are interested in their job. You don't need to work with the best of the best to work with great people; that's not what it's about. Working with great people is working with people you can approach with questions. People who help you, even if it means pausing their work. Great people are happy, smile, and want to be there. It makes an enormous difference working with great people. I've met many people during my different jobs, and I have kept in touch with at least one person in each job. This has been invaluable later down the line when they or I need help. It's remarkable how many friends you can make from working with cool people. It changes the way you see your work when you are having fun. I always say, 'Give me a great team and a horrible project over a great project and a horrible team.' It's that important to work with great people. The team is more important than any project you work on. Management can kick and scream about not hitting deadlines and having unrealistic expectations, but work is a breeze if you have a cohesive team. It's of greater importance to band together as a team, a united front, than to cave to management pressures. At the end of the day, management needs you more than you need them.

What Makes a Great Team?

A great team is self-organising, gels together, respects each other and helps others as much as possible. A great team is all about the individuals working as one unit. A team is only as strong as its weakest link, so it can ruin everything even if one person doesn't want to be there. Getting anyone who doesn't want to be there out as fast as possible is imperative. Their negativity and demeanour can hurt the team more than anything else. For this reason, you see high-performing teams happy and low-performing teams sad, frustrated and angry.

In the previous example of the company that made millions but had a high turnover rate, every person, besides one, was disgruntled, and as a result, the teams were miserable. However, management didn't care because they were making money hand over fist.

Miserable teams produce miserable results, and it's for this reason that you should leave as soon as possible if you find yourself in this situation. In this example, the only reason the company made so much money despite the miserable teams was that it was a product that required little in the way of advancement. This meant that the work done 15 years ago was still making the company a tonne of money.

A great team respects all the individuals in it. No one is too weird, too tall or too funny looking. Everyone respects each other, you don't have to love everyone in the team, but you must respect them and work together. A great team has a low turnover rate. Thus, they won't have to spend time re-gelling because they are in a constant flow state. If you were to look at any high-performing sporting teams around the globe, you would see that the cohesion of the individual pieces working in harmony wins the most over the longest time.

Take, for example, the 2014 championship-winning San Antonio Spurs of the NBA. They defeated, convincingly, I might add, the Miami Heat, who had a core of three superstars to the Spurs' aging former greats. It was a battle of the old vs new, with the old having played together for over a decade. The Spurs were known for their incredible team play. The ball would whip around the court, never sticking to one player's hands. Despite having an aging team and one rising young player, they took care of the Miami Heat with three of the best players in the league at the time, all of whom were in their prime.

This often happens; a team might have some of the best players but fail to make it very far. A team could have injuries, conflict, or not work well together and thus have an awful season. The teams that don't rely heavily on one player and have less-talented players fare much better. When their best player gets injured, the team is rather successful. There is someone who can take that player's place, slot in, and despite them not possessing the same level of skill, the team as a whole doesn't suffer. It is the same when it comes to all teams.

In the NBA, countless teams exceed expectations in a season, not because they are poor but because they don't have a superstar to rely on, yet the team fits together like a puzzle. You often see players who play as if they have never seen a basketball, but when a teammate takes a chance on them and puts them in a situation to succeed, with the right mentality and the right role to fill, they thrive.

This is the same with software engineers. Often people are placed in the wrong teams, technologies or projects, and they suffer. Then they move to the right company, or the right team, or the right project, and they thrive. People don't set out to be horrible at their job; they're simply not in the right situation.

The teams with excellent cohesion work well even when the talent isn't there. They exceed expectations because their output is much greater than the sum of their parts. They produce at a high rate because they work well as a team. For this reason, working in great teams is important, as it allows you to be a part of some interesting things. I have been a part of teams that might not have the most members or the most experienced members, but as a result of enjoying the work and team, they produce some remarkable results. Working in these types of teams is endlessly rewarding, and looking back, this is when I had the most fun in my career. The teams that are a joy to work with are not 50-people teams; they are single-digit-sized teams. They can work in silos for the most part and don't have endless dependencies.

One of the biggest killers of a team is being blocked by other teams, as it creates animosity between teams. I once ran a team building an application at the top of the stack; we had a direct user-facing application. Underneath, we talked to other systems, which meant we constantly had to reach out to other teams when things went wrong. We had daily outages that would halt our progress right near crunch time. These environments would be down for hours a day, every day of the week. This was demoralising for the team! You would constantly find yourself in a situation where things didn't work, and you had no control over it. It is difficult if you find yourself in one of these teams, but it's up to you to figure out how to improve it. We implemented a mock environment for all our external dependencies so we could continue our work with as little interruption as possible. We built our entire user interface from the mocked environment, drastically reducing the time reliant on dependencies. You find ways to work around it, and if you can't, raise it to the higher-ups. We were still a great team and finished as much of the work as possible in the given time frame and all the hardships we faced. We kept strong, made sure no one was facing too many hardships and supported each other as much as we could. Given the circumstances, we produced the best quality we could and were proud of what we produced.

SECTION 3

Master Your Craft

GOALS

I don't strictly write down my goals, and maybe I should, but I'm not here to tell you to do something unless I, at the least, do it myself.

Goal setting is vital; without it, you stagnate. I could sit back and clock in and clock out each day for the next 40 years. I could gradually improve (until I stagnate), use the same technology each day and progress as anyone would, doing the same thing each day. Hell, maybe I will even become proficient at my job. Yet, there's something incredibly depressing about that whole situation. I wouldn't have any aims other than to get in at 9 am and leave at 5 pm. My life would slowly waste away, and time would slip by like water through my hands, each moment passing but nothing sticking. It's this concept that I often think about as I head home from work, thinking that it's already Thursday, yet it feels like yesterday was Sunday. The workday flies by, and before you know it, you're in November, wondering where the year went.

I find that stopping and thinking about my goals and how I'm progressing towards them slows down the passage of time. It's like having extra webbing in your hands; it captures a little more time. For me, time is an elusive beast, but when I sit and plan, at least in my head, where I'm heading, what my goals are and how I'm going to achieve them, it gives me pause, the awareness of time. I hope it will stop me from waking up at 60, wondering where my life went.

Goals don't have to be written down, as every self-help book says. I don't write my goals down, but maybe I should; perhaps I'd get more focused, but I also know my goals aren't set in stone. I don't necessarily want millions in the bank or a Lamborghini in the garage or to be the best programmer alive. Furthermore, I just want to improve each day, getting that 1% better through directed self-improvement. I don't necessarily have grandiose dreams of creating the next Apple or Microsoft. However, I take each day as it comes because I find that opportunities come knocking if I leave myself open.

Stagnation

Stagnation is one of those states I fear most. As a goal-oriented person, I struggle with the monotony of daily life, yet I thrive on the routine it gives me. The routine is what makes it easier for me to function. I cannot drop a night at home to go out partying because someone asked me last minute. I also struggle with not doing something every minute; I feel like I'm wasting my time playing video games when I could be programming. However, I think if you have these goals set in your mind or written down, it pushes you to do the work, but it also gives you pause to slow down in daily life, knowing that you're heading in the right direction. It allows you to take time off and spend it with your family because you have a clear idea about where you're heading. Actively taking the time to rest and recover and scheduling it is much more rewarding when you know it's just a break to get back to your goals harder tomorrow.

Reflecting on how far you've come and where you still want to go is one way to hold onto more of that time that keeps slipping away. Reflecting on the improvement or the sheer volume of work you have produced is important because you often forget just how much you have done. You forget that just six months ago, you were here, but now you're over there. Unless you take this time to reflect on your goals and the progression you have achieved, you continue to lose time and let it slip by.

What you want in life is entirely up to you. You might want billions of dollars, or you might want to wake up each morning and create beautiful paintings. Whatever you want, you need to set goals for getting to that state. It's difficult to become a billionaire, much like it's difficult to become a world-class painter or even a decent painter. I believe that it's the aim and the constant practice of your craft that makes life rewarding. Taking each day to deliberately move towards your goals and reflect on your progression is what slows down the passing of time. We need to spend more time self-reflecting, thinking about the present and the future, and less time on our phones or watching TV. Most people lose weeks, months, and years of their lives because they don't take the time to set goals, reflect, and postulate on ideas, problems, and opportunities around them. Instead, they drown out the noise in their head with as much distraction as possible.

I'm guilty of all this myself. However, when I take the time to slow down and spend some time thinking, I enjoy life better, and I feel more relaxed and at ease. Life isn't about pushing as hard as you can but finding the direction you want to go and heading for it. If the direction changes, that is okay. Choose what new direction to take and take it.

Setting goals is one way to slow the passing of time.

How Do You Set Goals?

It doesn't have to be too formal; it can be as informal as writing your goals down on paper.

Pick one career, one life, and one financial goal. Give that goal a time frame, give it clear criteria to be met, and then regularly write it out or meditate on it. It's important to periodically review these goals to keep your mind focused on them. You can't achieve these goals if you don't spend time thinking about them, instead forgetting they even exist.

Take, for example, the goal of becoming a senior engineer by the end of your third year. This is somewhat vague as it depends on the company and the position they offer; not all jobs are created equal. However, this still gives you a pretty good goal because the senior title is usually clearly defined in a job posting or a title increase. You can then use this as your future vision to be that senior developer. You have given it a time frame and an outcome. Furthermore, you must regularly envision this future in which you are a senior developer. Once your mind sees and acts as though you are a senior developer, the world opens its doors to you. The hard work begins to pay off as people notice you are more senior, and you begin to progress faster. Keeping that clear picture of your goal in your mind's eye might sound farfetched, but it tricks your mind into believing that it has already happened. Once your mind believes that it has already happened, your body responds, and you become that senior developer well before you get the official title, and thus you become a senior developer in title sooner than you think.

Focus on one goal from each aspect of life at a time, making the focus easier and not allowing you to spread yourself too thin. Of course, you still need to do the work; you can't just think you are a senior developer without working toward that goal. You must put in consistent work, which becomes easier the more you focus on your goal. You will find it easier to work towards your goal when you have something to work towards.

Share Your Goals

Make others aware of your intentions; they will help you. You might be after a job in a specific sector, and it just so happens your friend knows someone who is hiring in that sector for senior software engineers. You get the interview and the job from that connection and goal-sharing. Likewise, you must share these goals to put them out to others to help you. Others will share their goals with you, and you must help them if you have the means. It can be as simple as offering a potential interview. You got breaks from other people's kindness; it's up to you to pass those breaks onto someone else. You might be one text message away from getting someone their first or dream development job.

Keep Aiming Higher

You should always have a goal in mind, something to strive for and keep going for. You don't have to push for a bigger title or more money. It could be as simple as expanding your skill set into a new area to work in a new sector. You might have spent a decade working in web development, and suddenly, you have an urge to work with AI. You could spend a few months training and go for a job in that sector. This would give you an entirely different set of problems to solve, and now you might have created 10 more years of interesting work. You don't have to limit these goals to just career ones. You might focus on your career, hit a point where you top out salary and title-wise, and then cruise. Now it's time to focus your energy on something else. This concept is explored in the book by Cal Newport, *So Good They Can't Ignore You.* Cal believes that you should not follow your passion. In fact, you should follow what you are good at, build that as a set of skills that are valuable to the market, and then create the life you wish to lead, leveraging the power of your skill set. For example, you get out of university, find programming moderately enjoyable, and are average at it. You get a job as a software engineer, you progress at a reasonable rate, and enjoy your daily work. It's not life-changing work; some days suck, some days are great, and the people you work with are great. You get 10 years into your career and find yourself wanting to pursue something different and change up your life. Perhaps you enjoy mentoring developers in your team and wish to do that full-time, without the pressures of deadlines and pushing some product out the door. With this 10 years of experience and five leading teams, mentoring developers and building people up, you decide to go out on your own and mentor developers full time. You set up a training business and get a few people through the door. It's not enough to pay the bills at first, but you have the development skills you can fall back on and the resume to show you know what you are doing. In this transition period, you have the power to keep contracting whilst building up your mentoring business, and you have that career capital. You have the resume and the financial safety net to pursue your goals.

Contrast this with when you are back at university: you know you want a job in the software engineering field, but you're not sure in what capacity. Just take the job in the field; it might not be your ultimate passion, but you will find that building a set of skills will add meaning to your career. In addition, you will have this set of skills to fall back on if everything goes wrong.

Cal talks about how most people never find their one true passion, and that's okay because it's overrated and often harmful. People are led to believe that their ultimate passion is out there, but it might not be. You might enjoy work, but it's still a job. You might enjoy painting Warhammer figurines, but it won't pay the bills. However, working three days a week or six months a year will allow you hundreds of hours for these hobbies. You don't need to make money from everything you do; you just need money to pay your bills, and then you can choose what you want to spend your time on.

Personally, I don't want to be a developer my whole life. I love development and getting so fixated I forget to eat, but I still don't want to do it all the time. I have many interests, and life isn't all about how you can be the best developer and earn the most money. It's about selecting what you want to do and aiming for it. I consider myself extraordinarily lucky in life, but not everyone gets this lucky. What you can do is plan your career to reach whatever goals you wish to achieve, and no matter your background, you will arrive there eventually.

ASK QUESTIONS

Kids always ask questions about everything and anything. 'How do lights work?' 'Why do I have to go to school?' 'What is that dog doing to that other dog?'

Questions make up a significant part of who we are. They allow us to wonder, think, and grow. They are what push us further in our mental growth. We ask millions of questions when we are young, but we ask less and less over time. Why do we ask so few questions as we get older? I partly blame the school system; its rigidity and lack of creativity discourage questions and encourage following formulas and not questioning anything. Furthermore, social constructs in schools discourage questions for fear of getting laughed at. The further we go with traditional schooling, the less we question things.

By the time you enter the workforce, you're so unequivocally unquestioning that you simply don't ask anything. Couple that with the internal pressure of needing to know and not looking stupid. Your boss could mention MVC when talking about web applications, and even though you might not know what MVC is, you nod and pretend you do. Then it comes time to do the work, and you panic because you don't know what to do. This whole situation can be avoided by asking simple questions.

Don't be afraid to ask questions; it's less scary than you think, and you just might learn a thing or two.

When I was in university trying to work out IP addresses for a networking subject, I just sat there and didn't ask any questions. I was too afraid to ask any questions because everyone around me seemed to know what they were doing. When it was my turn to write the next IP address in the range on the board, I didn't know what it was. I stood there with the classic hand-to-chin thinking position, trying to play it cool. After a few moments, I gave up, said I didn't know and asked what the answer was. Instead of everyone laughing, the tutor took me through it step by step. I might have looked like an idiot, but I learned something.

It's the same when you're working. You will frequently find that others don't know much, either. In large-scale organisations, there is a tendency to use numerous acronyms or team names, which make no sense to an outsider. It just takes one person to ask, and the group becomes more knowledgeable.

Asking questions is crucial; you won't know everything and never will. What helped me overcome my fear of asking questions is the notion that everyone knows something different. Now, this might seem obvious, but it helps. Allow me to explain. You've gone through your life; everything you have done has led to this moment, and every experience has taught you something. Conversely, your colleague has had an entirely different life; every experience has led them to where they are now. You might both be software engineers with five years of experience, and both went to Harvard, but you know stuff they don't, and they know stuff you don't. I can promise you that the things you know to be 'common knowledge' are not; we all learn bits and pieces. We all have gaps in our knowledge, so we must overcome our fear of asking questions, no matter how obvious they might seem.

When I started working, I always asked questions, no matter how stupid I thought they were. I've also been lucky to have some experienced developers on the team of my first project. I could ask everything and anything, even too many questions. However, in the 12 months of the project, I learnt more than I did in three years at university. Having someone to work with that allowed me to ask anything that came to my head helped my skills skyrocket.

Much like the curve on a linear graph, your progression will trend upwards in relation to how many questions you ask. You must ask questions as soon as you don't understand something.

The more questions you ask, the more you become comfortable asking them, and the more you ask, the more you learn. I feel like a broken record, but I will say it again: ask questions. Do not remain silent; silence breeds ignorance.

If you want to be the best software engineer, you never stop asking questions. Like Plato or Aristotle of the ancient era, you must never stop asking questions. Questioning things is important, understanding things is important, learning is important, and if you ever stop learning, your career will stagnate. You will stagnate. How could you not? If you're not growing, you're stagnating; if you're not stagnating, you're declining. The biggest companies get to where they are through constantly improving, then so many of them throw it away because they take their foot off the accelerator. One of the reasons Jeff Bezos is so successful with Amazon is that he constantly pushes the company to be better. This constant growth has propelled him to be one of the richest people in the world and Amazon as one of the biggest by market cap in the world.

If you want to improve every day, ask as many questions as possible and remember that everyone has something to teach.

TAKE RESPONSIBILITY

You get told to take responsibility for your actions as a kid. When you break something, you lie to your mother's face, even though she saw you do it. Your young mind thinks you can get away with it by blaming the dog or a sibling. Take responsibility; don't lie.

Taking responsibility is something many people forget about as they progress through their careers, especially as a junior. You're under the impression that a mistake is something bad, something inconceivable as a professional. There's no room for failure, especially when you start and even more so the further you go, meaning you can never fail. That's how it feels when you start, and that's how I felt when I started. I needed to be a perfect developer, who always knew what he was doing, always had the answers, and never introduced bugs into production. Well, I've done all that, and I've still got a job, and it doesn't make me any less of a professional software engineer. In fact, the more you make these kinds of mistakes, the more you learn. I broke a production database in my first year on the job by introducing a migration that wasn't properly tested and then pushed it to production, only to have a client pull out his hair because the customers were ringing him up asking what was going on. I freaked out and, for good reason, broke production to the point that customers could not use the product. It took hours to fix, and thanks to a more senior developer pairing with me, we fixed it. I took responsibility for fixing it, owning up to my mistakes, learning and moving on.

There have been many other times when I've made mistakes, but I've made a habit of taking responsibility for anything I produce. I could deny ever doing it, blame someone else, deny it's even a problem or blame the gods, but more than likely, someone will find out anyway. It will cause a lot more harm to your reputation when someone finds out that you're the person responsible for continuously breaking things and not fixing them, much like that kid in kindergarten who comes around and smashes your Lego right as you're completing it, only to run away and deny it when you confront them or talk to a teacher. These are the people you hate to work with; the lone ranger who goes off and rewrites a bunch of code, then commits it, and it breaks everything. The person who then blames the previous developers for writing terrible code, for he was only fixing that of lesser developers. The same person who fights tooth and nail for a project direction, only to hightail it when it goes south.

One sign of a great leader is someone who can take responsibility when they make a mistake, and when the leader succeeds, the leader praises the team; when the team fails, the leader blames themselves. The team can only be as good as the sum of its parts, but the leader is the multiplier on those parts. To become a truly inspiring leader, you must take responsibility for becoming a great teammate. You must take responsibility for everything you do.

I was contracted by a company that needed to deliver a white-labelled product. The problem was that we relied heavily on other teams to get their work done because we were the customer-facing application. We had deadlines for these teams, but they always missed them, and the teams were extremely unstable. We chose to start our development on their systems during their testing because we simply did not have time to wait until they had finished testing. We agreed that whenever we found an issue, we would raise it, and they would fix it. For months, we integrated with their systems, running into issue after issue, but there were issues even after they had completed their testing. We would constantly raise tickets for defects, and then the blaming began. Each team would blame the other, and this would happen constantly. Now, in fairness, all these teams had crushing deadlines that were impossible to hit, which just made everything worse. This resulted in us having argument after argument, trying to find out who was responsible. Dozens of teams could have been responsible as everything was a microservice (many small teams with many small codebases). Sure enough, we didn't hit our code complete. We took responsibility for the bugs and the missed requirements on our end; however, the teams underneath us that caused constant issues never took responsibility. This was all compounded by the fact that the environments we talked to, the ones we required to be up at all times to do almost anything, were constantly down. We had issues at every stage with teams pushing code into environments that broke all manner of things. I'm not putting hyperbole on this; it was constant. We had hundreds of raised defect tickets, all of which partially or entirely blocked us. Other teams would fail to test before they pushed, and no one would take responsibility for causing dozens of developers to lose time and projects to slip. The company was often advised to fix these processes, but they never did, and it always came back to bite them, costing money, time and broken promises to their clients.

Taking Responsibility Is Hard

Taking responsibility is hard. It takes work and can come back to bite you if you don't do it well. Bad companies and bosses are horrible to work for, and you get punished when you take responsibility instead of shifting the blame to others. However, the leadership it takes to take responsibility in front of your team is extraordinary, and you will get a lot of respect from your team if you do. Conversely, you won't get respect from constantly blaming other teams and people. The people you look up to most will be the people who take responsibility for everything they produce.

Taking responsibility is the key to having a long and successful career, so start doing it. You will gain tremendous respect, and it can be done at any stage in your career.

Delivering a Project

Start-up

In my first year, I was part of something that not many first-year software engineers get to do, and that was delivering a project from scratch to production. You dream of doing this from the first time you say you want to be a developer, but the typical entry point to your career is to start debugging some enterprise code in Java or some other enterprise language.

Before getting my first job, I applied for a job where I would have been doing enterprise Java every day. I'm delighted I didn't even get an interview because I was 'stuck' on an Elixir/React project in the education space. We took the idea presented to us by the two founders, contemplated what technologies we should use, set up an MVP goal, and had a release date—all that good stuff. We spent the first six months developing the product to release for a strict deadline. Much like all projects, it didn't hit that date, which was okay because, as you find out in projects, just because you said it has to be done by this date doesn't mean it's the perfect time. As both parties were new to the education space, we discovered that the deadline didn't have to be until about November, which was great because we needed that time.

We ran the project in an agile manner, running fortnight-long sprints. This allowed us to set out the tasks we wanted to complete, and at the end of it, we would have a retro, discussing what went well and poorly and how we would fix it in the future.

When you're part of a team in a big enterprise, for the most part, you work in a small team, which is one of many inside a large team. Your team will be sectioned to a very specific part of an application. You might be part of the payments team, where it's your job to create, debug and maintain all the functionality related to payments. That's it; you won't handle other features like accounts or other features. This is fantastic for when you start out. You're going to be in a team with other juniors, most certainly some seniors, and there's probably a lead at the top who communicates with the other leads from the other feature teams. This structure is one of many that an enterprise could have, but when you work on a start-up, you don't have the luxury of money and large teams. Each team member must wear many hats, and everything is done at speed, with corners taken in as easy to recover manner as possible.

For instance, our initial automatic deployment was only semi-automatic, and it was messy. It took a long time to complete, but we weren't DevOps engineers, which made it harder. That's what happens in start-ups; you do enough to get features over the line because you have little money and time, and you need these features for people to start paying for them. The pressure was high but not intolerable, which meant we had the drive to produce some great work without the fear that we would be punished if we didn't create perfection. As a team, we pushed hard, hit our goals, and were extremely proud of what we produced. Each team member learnt a lot about the technologies they were working on. For me, I had to pick up Elixir, a functional language. I quickly went from knowing nothing about the language and functional languages to knowing a great deal with a piece of software in production. I made countless mistakes, but I sat right on that edge, the precipice of being pushed just hard enough to grow but not enough to melt down. This is where the real magic of growth happens!

Enterprise

Delivering a project in an enterprise is the opposite of a start-up. For a start, there are thousands of rules. Secondly, you must deal with an incalculable number of dependencies. Most enterprises work with microservices, which are many small teams with small codebases or have many applications and thus many teams and people. It's just the nature of big business that there are so many people to deal with and many teams to interface with. In an enterprise, you spend 90% of your time in meetings, dealing with interdependencies and red tape. The actual programming part is the easy part; it's dead simple and takes almost no time at all. The other teams that you must interface with have deadlines and pressures, which affects you because, at the end of the day, they don't care if you're delayed, only that they make their deadlines and don't get yelled at by their boss. This creates enormous friction between teams, and the worse this gets, the slower everything moves. Next are the meetings; these are endless because every little decision requires a meeting, no matter how insignificant. Meeting after meeting, and the higher up the ladder you go, the more meetings you attend. This is one of the worst parts about enterprises. As a developer, you might get to skip out on a few, but the further you go in your career, the more you get into and the less you can skip. Even still, in enterprise land, when trying to deliver a project, you will have many meetings and waste a lot of time.

The final issue is red tape. Every enterprise has this, and the larger the company, the worse it is. If you're in something financial, it will be 10 times worse again because everything must be approved by dozens of people. In fact, quite often, the project is finished (code complete), and then it goes through months of testing, from performance, security, and accessibility; you name it, and they test it, and there is a team and meeting for it. In a start-up, you would do your testing, fix any bugs and then deploy to production. This could take a week or a few weeks or be done as you go. Over in enterprise land and more so in the financial sectors, this will take a lifetime. To be fair, there is a lot at stake; just know that the two worlds are complete opposites. I find that the most frustration comes from waiting on other teams to do testing or waiting for teams to finish what they are doing before you can implement the next feature. This is extremely frustrating as a developer, often leading to burnout, dissatisfaction, and people heading out the door.

You learn a lot when you deliver a start-up project because you have a working knowledge of every aspect of the project. You take part in aspects you know nothing about, so you must get up to speed rapidly. Furthermore, you come out of the experience as a much better programmer. Be warned; you can get into bad habits because you don't get the kind of quality you would have at an enterprise project; you simply don't have the time or the money to do so. However, that doesn't mean you will get this at enterprise companies; it ultimately depends on where you are. You can get quality from enterprise development by having the appropriate culture and the right mix of seniors and juniors. If you just have several developers who aren't willing to share their knowledge, you won't create a culture of progress. However, if you have many seniors who want to help others and you promote the practice of sharing, you get juniors that can create quality code.

I learnt a great deal from being part of a team that took an idea and bought it to life. I learnt that in a start-up, you must wear many hats. Your juniors take on much more responsibility than you could imagine. You learn every aspect of the code, the business, deployments, and customers. You learn from many mistakes and iterate on them rapidly when you go to the market. Delivering a start-up project is VERY different from an enterprise, but I love it. You get your hands dirty with all aspects of the project and take all the blame when it goes wrong, but hopefully, you learn from those mistakes. You get plenty of chances to fail, and you also get plenty of chances to succeed. I learnt more in my first year at a start-up than in 10 years at an enterprise. It was tough, but I wouldn't change a thing. They say it's not about the years worked but the quality of those years. I might be a little biased, but my first year feels like I learnt that of ten.

Delivering a project was extremely rewarding. Fixing bugs every day pales in comparison to delivering a project to users. Seeing the code you wrote being used by people in the real world or seeing a project go from an idea to development to production is extremely rewarding. The only thing that trumps it is the feeling of accomplishment when you finish a ticket you have been stuck on for days.

No matter what sized company you're at, you won't get over the feeling of delivering your first project. It's an intoxicating experience, one that I won't ever forget. Delivering projects is often the most interesting part of software development, and it shows with people always wanting to join start-ups to build the next big thing. This is where the most fun is, but it's also the most stressful. Hitting a deadline is extraordinarily hard, especially when the deadlines are almost impossible to hit. The key is to slow down, quiet your mind and relax. You will get there; it just takes time.

BE CAREFUL WHAT COMPANY YOU CHOOSE

You wouldn't believe the difference a company can make to your entire life. You spend a significant portion of your week at work, and if that place is miserable, you will be miserable too. I have worked at good and bad places in my time, and I can tell you I will take a significant pay cut to go to a great company over more money for an awful company. It makes that much of a difference. Take it from me because I've worked at some pretty terrible places.

Good Companies

A good company:

- Treats you with respect
- Pays you at or above market rate
- Gives you perks like an office budget or a learning budget
- Gives you opportunities to grow
- Conducts regular reviews with a system that rewards you for hardworking and hitting clearly defined goals
- Has great people working there
- There is no pressure to work more than 40 hours per week
- You are encouraged to speak up if there are issues and there are people to raise these issues with

A good company is a place where you see yourself working long-term. There might be bad days, but for the most part, the days are positive; you leave work feeling happy, healthy and not too stressed. You have access to resources around you in people and learning pathways to improve yourself. The company strives to improve its workforce rather than hire externally and let the existing people leave. When you go for a promotion, it is clear why you met or did not meet the requirements for the promotion, and you are provided with clear and measurable objectives to hit that goal. The company supports and pushes for goal setting, and if you want to progress in your career at a rapid pace, it has the structures and systems to get this done. Not all companies have these structures, so if you intend to progress rapidly, you will quickly outpace the company, its salaries, and its positions. If you do this, leave immediately, not because the company is bad but because they simply aren't at the scale to support you.

The company should have interesting work or allow you to move between projects, so you don't get bored with one project. I highly recommend never staying at a company or a single project for more than a year unless you are still getting valuable learning from it. This is your career, and if you are not getting what you want from your place of employment, don't feel bad. You can talk to the manager, and if it can't be resolved, leave for another company. A good company will try to get you to stay. They will push for people to be put into positions that will stretch their skills or move them from one project to another if they have been there too long. A good company will follow clean code, test appropriately, release code often and resolve issues quickly. A good company will not stand for average code but push for excellent code and the deadlines to allow for it. You will know when you are at a good company because your life will be easier, and there is a support network around you no matter the stage in your career.

Bad Companies

A bad company:

- Pays poorly or pays huge amounts just to keep you there
- Has a gigantic proportion of contractors to permanent employees (this can be a big red flag)
- Does not have a clear path for promotions
- Does not have clear goals
- Does not have learning pathways or budgets to learn on your own
- Does not have perks
- Installs surveillance software on your computer (this is a massive red flag)
- Has constant deadline pressures
- Management never lets the development team fix up code after rushing to hit a deadline
- Does not do testing
- Has culture issues
- Does not reprimand unprofessional behaviour
- Does not condone inappropriate discussions in the workplace
- Asks you to work overtime with no extra pay or pressures you into it
- Has managers or owners who act inappropriately in the workplace

A bad company will have unrealistic deadlines; thus, code will get rushed through without tests. There will be manual QA, if at all, and hundreds, if not thousands, of bugs. You will be under constant pressure to get your code out the door, with each sprint having more tickets than hours in a sprint. You will work overtime and on weekends whilst your stress levels go through the roof. You hate each day at your job, and as a result, your life will be miserable. You never get time to learn anything new, apart from being forced to upskill as quickly as possible to get the mountain of tickets done and out the door. When you put in 80-hour workweeks for months and ask for a raise, they either laugh you out the door or give you a 2% raise (this happened to me). These companies never promote you or raise your salary to anywhere near what you deserve. They continue to work you to the bone, and as a result, they have a high turnover rate, with most people not making it past one year. A bad company threatens you to work overtime to meet impossible deadlines they set for unknown reasons. You never see the goal; you only see the deadline. I have been at a consultancy that made it clear that if we missed the impossible deadline they set, the client would walk. If a company is putting the pressure of a client on its developers, it is time to walk. As a software engineer, you should never feel the external pressure of a client; that is what your tech lead and management team are there for.

How to Tell What Kind of Company It Will Be

During the interview process, how a company will act can be clear. It can be harder with some companies and easier with others, but there are a few guidelines on things to look out for. If you see more than two or three of these in one interview process, you should turn the job down if you get it. Going to one of these places is not worth your sanity unless you desperately need the job. If you can afford to be patient, be patient, as it will help you eventually. The worst thing you can do is go to a terrible company and be completely put off the entire industry. I promise not all places are like this, only the bad ones.

What Should I Look For in the Interview Process?

Rushing to Schedule the Interviews, Being Inflexible with Timing

You have commitments or might have a job, so you can't just drop everything to travel to the city for an interview. They must be flexible; if they're not, don't give them the time of day.

Poor Interview Process

This could mean many steps, including jumping through hoops to get the position. Note: big companies like Google or Atlassian will have very long interview processes, but this is normal.

Stressful interview

If they trigger your stress response in the interview, it will only worsen. The following are common interview stressors for bad companies:

- They make the coding challenge too hard, or the tooling is awful. I once had to perform a coding test in front of five developers in a Google Doc with instructions from a man with a heavy accent. I did not get the job; I didn't even understand what he was asking me to do!
- The coding test is too long. I've spent eight hours on one coding test only to not finish and not get the job. If you can't get through the test to about 80% or more in one hour, stop and tell them. You should never spend more than one hour on a test unless they pay you, which they won't.
- The people interviewing you don't seem like good people to work with. There is a high likelihood that the people you are interviewed by will be the ones working

alongside you. Often, you will be interviewed by the tech lead of the team. Don't take the job if they appear to be a weak leader or not technically strong. You just missed out on spending 40 hours a week with a bad boss. I had turned down a job before, and this was one of the red flags. The tech lead asked one question during the entire interview and left the rest up to the other non-technical people on the call. Not only did I have zero trust in his ability to lead, but he had no clear opinions on anything, which is not what you want in a lead.

Poor Contracts

There are some nasty contracts out there, so be cautious when you read them. Some say you can't do anything outside of work. Some say you must work more than 40 hours. Others tell you they have permission to see and do everything on your computer. I have also turned down a well-paying job because the contract stated that they could monitor everything I do under a surveillance clause. Now don't get me wrong, all workplaces have some monitoring, whether email, messaging, or internet. However, there is a difference between the internet traffic for everyone in the office, taking screenshots of everything you do every 30 seconds and logging the time you spend at your computer to the second. Any company that does this will be awful.

I have seen contracts with extremely restrictive non-compete clauses, meaning you can't do any work for 12 months in the same field after you leave. Now, depending on the country you are from, this might be illegal, unenforceable, or highly restrictive. Watch out for these clauses, as they can come back to bite you.

Other contracts attempt to lay claim to anything you produce. For example, you decide to make a website that sells shoes online. You do this entirely in your spare time on weekends and after hours. You release it to the public, and it goes swimmingly. Come Monday morning, you get a phone call from HR, and they lay claim to all your profits and the business because your contract states that they own everything you do. These contracts exist and, depending on the country, can be heavily enforced. Make sure you always get in writing all your questions with answers, especially when it comes to working outside your day-to-day job. A rule of thumb is that as long as it's unrelated, it should be all good, but check in your country and with a lawyer if you intend to make something of significance.

Saying One Thing and Doing Another

Companies often post a job listing with a salary range. Occasionally, they post it without a range, and that's wonderful; you know exactly what the salary is and everyone is gunning for the same number. What is unacceptable is being lowballed by 10% when you get the job offer. If they think you're almost at that mark but not quite, that's also okay, as long as they explain exactly why you didn't meet the mark and how you could meet it in the future.

Another red flag is when they say that the number would be higher than anyone else or higher than other people in the company. This is an instant walkaway for me, and it should be for you, too, because it means they don't pay their other employees well and won't give you pay increases either.

As mentioned, a great company will strongly persuade you and will explain exactly why you didn't meet the mark and how you could meet it in the future.

Pushy

This often happens more with the recruiter who handles the process. Most countries would have recruiters handling most of the talent acquisition. I once received a call from the recruiter handling my case, saying I got the job but needed to commit on the spot verbally. I told him I wouldn't because verbally agreeing is as good as signing. You need to review every contract with a fine-tooth comb, which is how I found the surveillance clause. I did not accept the job and told them exactly why.

A good company will always give you time to think about the offer. At a minimum, you should be allowed a couple of days. Any company that says you need to accept in a rush usually tries to hide something. Any company trying to get you to join without a contract is certainly hiding something.

Picking the Right Company for You

Choosing a company to work for is hard; you never truly know what they're like without working there or knowing someone who works there. This can make it incredibly hard to decide, especially when you have a job and like it. Is the grass greener on the other side? You will never know until you take the plunge. Conversely, you will never know what type of company you like until you try a few. You will certainly work at companies you hate and some you love; it's all part of your career. You might love enterprises or love start-ups, but you will never know until you work at both. I urge you to make sure that you pick companies that treat you well in the interview process, have fair contracts, pay you fairly and don't pressure you into anything. As mentioned earlier, these are marks of a great company.

Tech Lead: The First Step in Leadership

I love software development. There is something about building from the ground up, starting from nothing and creating something people use. The feeling of writing code is so intoxicating that I often forget to eat. In contrast, I have always been drawn to leading a team. I wanted to lead a team from my first day as a developer. I have been a part of many teams, so I have seen the good and the bad in teams and leaders.

Being a tech lead is something I always thought I would be good at because I thrived on mixing it up between development, delivering a project, and managing the team. I find it easy to pick up what drives people, their strengths and weaknesses, and what tasks they can tackle. I have been a tech lead on a project at IAG (Insurance Australia Group), a huge client and absolutely intimidating as my first lead position. Here is how it went, what I struggled with, what I did well and what I still have to improve. I write this as my first interface with being a tech lead, something I had been dreaming about since the beginning of my career.

My First Project as a Tech Lead

I got a call from my engineering manager one day, completely out of the blue. Usually, I have this fear that when that happens, I'm in trouble. I think my fear comes from always getting in trouble at school when I was a kid. My manager explained to me how I was needed at IAG for a project that was underway as some team restructuring and a tech lead position opened up. This was extremely unexpected but great news. However, my fear immediately set in. How would I manage multiple developers for a client with 15,000 employees? I had been a core member of many teams that delivered software from idea to production, but to be one of the people at the top was new. I have read many books on leadership, software, and biographies about great people who were great leaders. Still, I knew nothing and had no real experience aside from a few projects where I had to manage some people or train a junior here and there. So, I ferociously read and researched a few weeks before starting.

When I finally started, my fears were immediately stamped out because I was integrated into the team for a few weeks, and my team would start our project four weeks after I joined, giving me a few weeks to ease into the client, get my head around the project and start getting to know the team. My biggest fear was less about delivering the project or my leadership skills and more about my fears around not being at the 'tech lead' level in my development skills. However, this was not true, and I even surprised myself. It took a few weeks, but I could better understand the codebase and help most people in most areas. For the first month or so, I felt like I was drowning, never having enough time to answer questions, help the developers, understand the codebase, talk to higher-ups, etc. It felt like a constant game of treading water. However, it wasn't long before I got into the grove.

What I Did Well

People Management

This was something I wanted to focus on a lot. The better the team feels and the happier the team is, the higher quality of their work and the more fun the team has. I have been in teams where the leader bends over backwards for the client, and the developers pay the price. This is something I wanted to avoid. As developers, we know that sprints sometimes have infinite sprint points, which carry over from week to week. I was very vocal about sticking up for my developers, not overloading them and keeping sprints in check. If work changed or new requirements came in, it didn't simply get added to the sprint; it was put in future sprints or reprioritised.

I checked in with my developers regularly to find out if there was anything I could help with, either work-wise or personally. Every so often, they would highlight their lack of challenging tasks for the last few weeks, so I would prioritise certain tickets for certain people. If someone were struggling, I would get them to pair with another person. I quickly learnt what level everyone was at, which allowed me to divide tasks by skill level. When there were conflicts between team members, I managed to deal with them effectively, but this was the most challenging part of my role.

Context Switching

I have never context-switched so much in my entire life. Between three separate messaging platforms, email, pull requests, meetings, developers asking for help and trying to get tasks done, I was busy. I spent a lot of time switching between tasks. Again, this goes back to time management and keeping track of your tasks.

My method was simple. Each day, I would open a dated file, put all the tasks from the previous day in a list that didn't get completed, and begin adding tasks as they came throughout the day. If a task wasn't a high priority, I would worry about it later, but rather than forgetting about it, I kept a record that it existed. There was an instance where I had to provide some feedback on architecture for the client. This needed to be done but would not be completed that day, so it went on the list and was completed when I had some time a few days later. If I didn't have this list, I would have had to stop what I was doing, do it, send it through then jump back.

Context switching is almost impossible to avoid as a lead, but you can find ways to reduce it. Even with this list, I was still swamped with requests always coming in. I eventually enabled a constant 'do not disturb' feature on macOS to reduce the amount of 'pings' and 'pongs' from my computer.

Putting the Team First

My goal was always to put the team first, no matter the pressures of the client or the project. You don't have a project without a team, no matter how much you jump up and down. As much as possible, I would shield the team from the pressures of the project; thus, the team was happier and produced better work.

You can't always keep the team happy, but at least shielding them from all that direct pressure helps them work more efficiently.

I have been in the same position, where a project has enormous pressures on it from upper management. It's extraordinarily tough to get work done with someone above yelling in your ear to hurry. You work slower, you're more stressed, and your health suffers. If you're not careful, you lose people, which causes more harm than management realises. When I was a lead, there was a lot of pressure on the project to hit deadlines, which meant I would face those pressures directly. If I were to relay this pressure to the team, I would have instantly created resentment in the team, they would have worked less efficiently, and ultimately, they would have hated their jobs, the client and me.

The worst places to work are those where the pressures are passed from high to low. As a developer, I should know there is a deadline, but I should never feel the pressures, or at least not enough, to affect me.

When I was a lead, instead of putting the pressures I received and immediately putting them on the team, I would look at the sprint, reprioritise what we needed to do, and move people around if I thought this might benefit the delivery time. There were circumstances where dependencies were causing a host of issues. Instead of jumping up and down, yelling at my team to move faster despite tickets taking longer, I reprioritised these tickets so that people could work on other tickets that were not impacted by other teams and, instead, focus our team on the tasks within our control. I would then follow up with these teams on the blockers. Keeping morale high, keeping the team happy and keeping the goal plain and simple are all vitally important.

Dynamically Handling Blockers

In large companies, there are many teams and deadlines. Each team has its set of dependencies and deadlines. The closer you are to a user-facing application, the more you rely on other teams. If they haven't finished their work due to one of their dependencies, then they are delayed, and you are delayed. We had a lot of this as we were on the top of the stack, a direct user-facing application on a brand-new platform. Often, we would have a sprint to integrate an API, but the problem was that this API was delayed, which meant we had to reprioritise. Other times, an API was there, but it wasn't tested, which meant there were bugs we would have to raise. Rather than having all my team sit on these tasks, I would divide them up so that some were on less critical work that wasn't blocked but still could progress on the more important work. You might wonder why I chose to work on the lower priority work, and the answer is simple: those APIs would often be buggy, go down or just not work, which resulted in many people trying to figure out whether it was what they were doing or what the API was doing. So, if you could complete all the work in your control and the remaining work that was reliant on other teams took longer, that was okay. You could raise that to the client, and it would be off your hands. We could only control what was in our sphere of influence and depended on other teams; thus, it was up to me to sort out the issues, not my developers. At the end of the day, if we didn't hit the deadline because my team couldn't get the work done, it was not the developer's fault; it was mine. It wasn't our fault if we couldn't hit the deadline because we relied on other teams. We did the best we could. I also never blamed the dependent teams because they had too much work to meet their deadlines.

What I Could Improve

Time Management

This was a different ball game to being a developer, and I think it was the nature of the client. Many days would be taken up by meetings, leaving very little to help developers or pick up tickets myself. I had to employ a personal task management list in which every task I had to do would go into my list, no matter how big or small. Then as I did them, I would check them off. It was incredibly basic, but it was a necessity. Otherwise, I would just lose time and wonder what I achieved every day. It was great to look back and see a list with the following; I reviewed these pull requests, helped this developer, and gave this information to some testing team. Still, as we got closer to release with more and more testing teams using our code, there were times when they would come to me with something broken. This would block their entire team from testing something, which meant I would need to rush a fix in or figure out the problem, so I could give them an idea of how long it would take. These would pop up daily in the early days of testing, as these things do. As a result, I quickly lost entire days trying to resolve these issues between meetings, reviewing code, and doing my work. It was less about the time to address these issues, as I could typically fix them quickly. The harder part was their sporadic and unpredictable nature. I could look at my to-do list in the morning and see a couple of items, then by 10 am, have two bugs block entire testing teams, and I had to fix them. So, if you don't keep track of your tasks, you quickly lose your days, wondering what you accomplished.

This holds true for communications between teams. As the technical lead, you are the communicator between your team and external teams. Furthermore, you often get pulled into meetings last minute to provide technical insight into things, so if you are not keeping track of what you need to do and what you have done, you quickly feel like you are drowning. It's all about being strict with prioritising work, as there is only a limited number of hours in each day.

Saying Yes to Far Too Many Things

As a senior developer, you are expected to take on the hard tasks. This might be setting up an integration testing framework with CI/CD or migrating to a new framework. I put my hand up for too many tasks, whether developing or getting something for another team. I would say yes too frequently, which meant that something for my team would get pushed back.

There was a time when I picked up a ticket that required a lot more work than I initially thought, which resulted in not having enough time to help my developers. However, when I started offloading these tasks to other developers, it freed up time for me to help others and do the more important things. I was not brought into the team to get hard tickets across the line; I was brought in to lead and get the team to complete the work before the deadline. As a lead, you must leave your ego at the door and let your team shine. It didn't matter who did the work; only that it got done efficiently. I became a better leader when I stopped taking on too much work. Driving a bus is hard when your eyes are not on the road.

Having the Tough Conversations

I am naturally conflict-avoidant. This meant that with so many team dependencies, most of which caused us delays in one form or another, there had to be hard conversations. I believe in being respectful, calm, and acting with empathy because everyone has deadlines, pressures, and issues. However, it was hard to have these hard conversations, and it was something that I needed to improve. For instance, I needed to get stronger in pulling people into conversations from other teams when the bugs were not getting fixed in a reasonable amount of time without fearing they would hate me for it.

Being the technical lead in a team is what I thought it would be in some ways and radically different in others. I loved many moments of it and hated others. Yet, I loved working in the team I had. They worked their butts off, handled the pressures of the client with grace, and were positive and gave their best every day.

I love being a tech lead and love delivering projects. It was a tough road, and I made mistakes, but I also accomplished a lot. The more I trusted my team and would get out of their way, the more they produced their best work.

I will end this with the most important thing I learnt—having empathy for everyone around you is the most indispensable skill. I highlight this specifically because there were times when you could jump up and down about why someone was not doing something or why the quality of something was poor. You would miss a lot if you didn't empathise with their situation. Developers often blame others for bad code, but the empathetic way to deal is to assume they simply don't know any better. People don't go out of their way to do a poor job, which leads me to conclude that they simply don't understand something. If you go in with an empathetic mind, you find that the code this person wrote isn't poor because they wanted to write poor code; it's poor because they didn't understand something. However, once they understand, their code skyrockets in quality. As a tech lead, empathy is truly invaluable.

LIFE CYCLE OF A PROJECT

If you're just entering your career, you might not know how code gets from idea to production. I will use start-ups as examples below, but you can still apply this to enterprises.

Idea

Bob comes to your company with an idea for a new GPS app. Your company 'Big App Ideas' (BAI) brings Bob in for a meeting to discuss this idea, the process they will follow and how much it's going to cost. Everyone is happy, and Bob signs a contract. This is the first phase.

Inception

Bob attends meetings with designers, product managers or delivery leads. This is where the main ideas for the application are fleshed out. Furthermore, BAI pulls out the app's most important features and plans a roadmap of how this will fit together. All the ideas Bob has in his head are thrown on paper and ranked from most crucial to least crucial. Through discovery, the biggest feature is the ability to use the GPS, so this is the first feature that is planned. As the hours drag on, the features get less and less essential, and ultimately, the MVP (Minimum Viable Product) is produced. This set of features must be completed for any user to value the software as something they would pay for. This typically means the smallest number of features possible without removing something vital. The list of features is there, a roadmap is created, and a timeline is set.

Design

Designers then spend time with Bob to get the look and feel of the application, such as its colours, based on the emotions and messages your application should convey to the user. This is a crucial process; it ties the feature set into something beautiful that draws the user in, capturing their use. No matter how good your application is, most people won't use it unless it's visually appealing. There is a reason design has evolved from the early 90s; you're less likely to use an application when you don't visually enjoy the experience.

Proof of Concept

Once the design is done, the proof of concept phase begins. This is where a designer takes the design and creates a prototype. This is then shown to the client, and if approved, it gets handed to a development team.

The proof of concept is different to a prototype. A prototype is typically directly lifted from the designs. Most design tools allow you to create an interactive application right from the designs. However, it doesn't do anything except look nice and move through the screens. The proof of concept is where the development team takes the main feature and builds it out. This involves finding what libraries exist to help, what tools are out there, and how this could be implemented. Once all this information has been gathered, a quick implementation is created. In this example, a phone might show the GPS screen being updated as you move around. This has almost no value to a user but proves that you can use the main feature. Furthermore, it proves the difficulty of getting this feature complete. This is key, you won't have a complete GPS after the proof of concept, but you have an idea about the difficulty involved. The difficulty is half the battle in software, as this is where you lose the most time, and you won't know this until you build it!

Development

The next phase is the development phase. This is where developers come in. Often, you work in sprints. As a team, you plan two weeks' worth of work, and you pick up these tasks for the next two weeks. You try to complete as many as possible, but you shouldn't be bringing in more than your team can handle. Ultimately, this is hard to determine, but the more you do it, the more the team understands how much can be done in each period. As the sprints tick over, you fall into a rhythm of work. You have weekly meetings to go through tickets, performing what we call scoping. This is where the ticket is read out, discussed, and finally, a rough timeframe is given. These estimates are given in many ways, mostly in story points (1,2,3,5,8,13).

- 1 point: less than half a day
- 2 points: up to a day
- 3 points: half a week
- 5 points: one week
- 8 points: up to two weeks (should be split up if possible)
- 13 points: greater than two weeks (must be split up into smaller tickets)

Some teams use different number formats; however, they all mean the same thing. It's a way to estimate how long something might take without being fixed to set hours.

The newer you are to a project, the harder this is and the more wrong you are. In other words, you get better and better over time; however, you will never fully nail it, but you'll get closer, and that's the main thing. During each sprint, you can determine how much velocity the team has by looking at the previous sprint to see how much was completed and left over. You can then bring the number of points you finished back in, minus the leftovers. You work in two-week sprints until the project is finished.

How Are Tickets Created?

Each ticket will be created by a BA (business analyst) who takes the requirements for a feature, usually from the client, and translates them into actionable items for a developer. If needed, designs will be provided by a designer and the acceptance criteria (AC). The ACs are crucial because they determine how a developer knows they are finished with the ticket. You must meet each one of these items to have the ticket deemed complete. Generally, they are written in the form of 'if this, then that.' For example, 'If I click on the signup button, I expect to be signed up and redirected to the home page.' If your code fails this, then the ticket is not finished. This provides clear guidelines on what you must achieve for this ticket to be marked complete. Furthermore, you require tests to ensure it's running in a hosted environment (this should already be done for you and be automatic). The ticket must also have a description of the feature and any background context.

In the example with a signup button, there would also be a linked ticket to be first completed that builds the signup form. This ticket would simply be to add the button that redirects you and sends the information. This is important because you need to understand the scope of your tickets so you can properly estimate them. You don't want massive tickets such as 'Build signup form' as this is much too big for a single ticket. You want to break them up into small pieces, then submit a Pull Request for each ticket, merging your changes to the overall project at regular intervals.

Release

Once all the work is completed, it's time to release it to some users. Within start-ups, you would release it as soon as possible with your MVP. You gather a select group of users that can test your application to see if they like it and provide feedback. You discover many bugs, and performance issues, which you fix in the coming weeks. Most importantly, you get an idea of whether people enjoy your software; there isn't much point in building it further if no one uses it. Now you must maintain a production set of code, meaning when you make changes, you have real people affected by it. This means that you can't break their experience, or you will hear all about it. If you break production, this significantly affects how your users see you and your product. Once your product is released, this drastically changes the game. No longer are you just building features; you're now building features and addressing problems. You are taking feedback on the product and making changes to the feature roadmap. It's incredible to see your code in production, working all these hours, and now you have people using what you created. This is a truly fantastic feeling, something you won't soon forget.

Maintenance

What happens next depends on the product. If all the features are complete, you might fall into maintenance mode for a while until there are some new features. This means fixing bugs from most critical to least critical. This isn't an exciting phase, but it's vital; all the shortcuts to meet a deadline are fixed, and the bugs are closed. You now oscillate between bug-fixing and new features, perhaps a 50/50 split, depending on how aggressive the new features must be. In other projects, you might go onto phase two, which is the same as the first; you just have bugs and user complaints to attend to and a gigantic feature set to complete.

To me, these projects are the most fun in programming because you get to take an idea to market and get real people to use it. The feeling of accomplishment you get when your hard work is out in the world is intoxicating.

PROFESSIONAL
SOFTWARE ENGINEER

What Is a Professional Software Engineer?

Software engineering is a funny field; it's complicated, and thus, you would expect some education requirements to practice it. Much like law or medicine, or other engineering fields, they all require people to pass specific requirements and meet standards. However, we don't have this in software engineering; thus, anyone can be allowed in as long as a company is willing to hire them. Therein lies the problem; most people won't have formal education or training, resulting in poor overall professionalism in the workplace. This does not mean that you need a university degree. No, the lack of time and experience results in people not knowing what they don't know. I regard university as something that most people don't need; however, the one thing it does do is that it forces you to commit to a set course for several years, meeting certain requirements along the way. University may teach you a lot about nothing and be very outdated. However, it produces people who have at least followed the path, which is essential. It raises the barrier of entry not to an elite level but to a minimum standard. You wouldn't want a random person working on the electrical wiring in your house. They might start a fire or burn your house down. Therefore, there are requirements to be an electrician.

What trades do best is 3-5 years of supervised-on-the-job training from an experienced tradesman, with minimal book learning. You have a certain path to follow with your coursework, which is mostly ticking boxes and understanding the basics.

The rest is learning the tools, which is what we should have for software engineering. I have said it many times, but you learn the most from doing and doing the most when you work full-time in a team of experienced developers. This is not the norm, so I digress. This is not to say you can't be a professional; it's just not taught well outside the university setting. If you don't come from the university path, and even if you do, you need to actively work at being a professional in the workplace. The nature of a 3-5-year university course does stress a greater level of importance when it comes to quality. Compared to a 12-week intensive boot camp, the skills you come out with are entirely different to that of university. From boot camp, you will understand modern technologies and how to put something together in a short period of time. In university, you spend years learning about theory, having the quality required to achieve great marks drilled into you. University doesn't do much right, but it produces people with a higher emphasis on quality.

What Is a Professional?

A professional is someone who does not cut corners, prides themselves on writing clean code, testing that code and working in a respectful and productive manner. A professional is someone you would love to have on your team because, regardless of anything that happens, they get the work done to a high standard and do not complain about it. A professional won't blame others for their mistakes; they take accountability for everything they do and won't settle for second-best. A professional produces high-quality work at home, in the office, or with 1000 eyes staring at their code. A professional is the ultimate workplace champion. A professional is someone you must have on your team if you want to deliver high-quality software that will last. Many businesses fail because they don't have professionals in their ranks.

Why Is It a Must?

When you think about all the people you could work with, would you want to work with someone who blames others, writes poor code and cuts corners? No, you also wouldn't want this in the builder building your house. You would want a professional who doesn't cut corners, who builds it watertight and to specification. This is the same in software development. You most likely won't put anyone at risk of death or injury; however, you should still treat your work as something to take pride in. The work you produce reflects who you are. If you produce shoddy code and cut corners, you won't be respected and will not progress far. Your career is all about your reputation. A good reputation takes you farther than you can imagine. In the past, my reputation allowed me to skip parts of an interview or the entire process. A good reputation takes a long time to build and can be destroyed in seconds. You must take pride in your work, if not for yourself, then for your team. You exist in an ecosystem of people all working to reach a goal. If one person constantly pulls everyone back, the team will begin to despise them. They won't be liked and will eventually be removed from the team. You must strive to be an effective and professional member of the team. Others around you drop to the lowest quality of the team, so it's your job to raise this as much as possible to set an example of what the team should be doing. The more you practice this professional skill, the better your career and easier your career becomes and the more respect and enjoyment you get out of your career. People will want to work with you and ask you to join their team on new projects. Your boss will put you in projects that are more interesting or have higher stakes for the company, where the action is! All these things come from being a professional, holding yourself accountable, and striving to improve daily. No matter your career background, it's never too late to start being more of a professional. Whether you have a university degree, are self-taught, changed careers, or did a short course, it doesn't matter. You can be unprofessional and professional from all backgrounds. It's up to you to be a professional, to carry yourself above mediocre standards and to produce work you are proud of.

How Do You Become a Professional?

It takes hard work and dedication, but anyone can do it. Where most people would simply push their code, you strive to have it well-tested and all edge cases sorted. You don't accept band-aid fixes; instead, you find and fix the root cause. You refactor code when you see it can be improved, and you always leave it better than you found it, for you know that code rots over time. Being well-read in the field also helps you be a professional. However, it mostly takes work, which most people don't put in! Most people are too lazy or don't care enough to do a good job; they just get the job done. At the end of the day, you can choose to become a professional, but it takes hard work. Your reward will be producing high-quality and well-tested code and having the reputation that you always output code this way. A reputation takes a long time to foster, but once you have a good reputation, you will be known as a professional who always produces great work. You will be selected for teams and jobs that are the most interesting or require the best people. This is where you want to go. This is where the professional path takes you!

Where Can You Go?

The sky is the limit when you are a professional. You stand above the rest, skipping steps in the interview process and getting the best jobs. Being a professional is the fast track to a great career, but it takes effort. Being a professional is something you must do every day of your career, but if you do it, you will go far. If you took two people with the same experience, you know who would get the job: the more professional person. It's all well and good to have delivered a project for a big company, but a professional will not only deliver the project but also lift others up. You might even take someone with a lesser resume if you think their influence on the team would have a greater positive impact than someone technically superior. This is the power of a professional attitude to your work. A team of professional programmers can move mountains; these are the elite of the elite when it comes to delivering projects. If you ever get a chance to work on one of these teams, this is a team to treasure.

The answer to where it can take you is that it can take you to the moon and back if you let it. The world lacks professional software engineers, so if you are one, you will stick out and claim all the work.

CLEAN CODE

What Is Clean Code?

Clean code is code that others can easily understand because, in its nature, everything is broken into small descriptive parts. Variables are adequately named, make sense, and convey their function's meaning. The same goes for functions; they aren't big and bulky, but they are broken into small reusable pieces that also convey functionality inside to the outside world.

Clean code isn't something you simply stumble into or must wait 10 years before you can do it. You have to actively fight the urge to produce poor code. There are deadlines and other people who output code faster, but you need to be professional, as Uncle Bob (Robert C. Martin) would say. The professional programmer strives to produce high-quality, clean code that contributes positively to the entire codebase, not slowly bringing it down to an unmaintainable mess.

The biggest problem with clean code is that it's hard to do. Each time you add code to the codebase, you are making your codebase larger and worse, its entropy the slow constant decay. You must actively fight this entropy, which makes it so hard. It's not enough to keep clean code in mind; you must actively pursue that goal each day. The Boy Scouts have a saying, 'Leave the campground cleaner than you found it.' This is precisely what you need to do as a professional. You must constantly improve the codebase to allow it to thrive, not decay over time.

The Core Tenants of Clean Code

Following the Conventions of the Project

Every project and language has a convention. These conventions are rules and guidelines for structuring, writing and interacting with the software. A professional programmer follows these at all times. It doesn't matter if you don't agree with how a language structures itself; you must deal with it and follow it. Nothing is worse than someone who chooses their own formatting for a project compared to the rest of the team. It's up to you to follow the principles of the language and what the team is doing as a whole.

Keep Your Code as Simple as Possible

Code complexity is the root of all evil. It produces monstrosities of code that no one understands or can change, and it causes bugs. The more complex your code is, the more potential for bugs and the harder it is to change your code when needed. You must write your code as simply as possible. It must be so simple that you can come back to it in a few months when you have forgotten all about it and not swear at the creator of it— you! The developer who prides themselves on complex code is a poor developer and awful to work with in a team. The developer who can make the simplest code possible is a great developer.

Refactor No Matter How Small a Tiny Part of the Area That You Touch to Improve the Codebase Overtime

Every time you pick up a ticket, refactor whatever is of poor quality in the codebase at the end. No matter the size, finding areas that could be improved and not waiting for someone else to fix them is good practice. It can be one function or a couple of variable names. But it doesn't matter what it is, only that it improves over time.

My greatest achievement in my career was getting into a codebase that was complex and hard to deal with. Another developer and I mapped out the biggest annoyances with the codebase, and we planned our small refactors to improve this over time. After six months of doing this, we left the company, and the codebase was spectacular. It was easy to refactor code, understand what was happening, and test. Our extensive test suite highlighted any issues when we refactored code; it was spectacular! You can implement the same process as you work by making small incremental improvements that don't amount to much when taken one at a time but, over a long period, generate tremendous benefits for yourself and the team.

Never Band-aid Fix a Problem; Address the Real Problem and Not the Symptom

You often come across a bug or a problem in your code that just doesn't work properly. You spend a long time trying to figure it out, but at the end of the day, you still can't. Don't quit and say it's too hard. Instead, keep going and reach out to others for help. You should never implement a band-aid fix unless there is an extremely valid reason. It's up to you as a professional to put in the work to implement a proper fix, not leave it for someone else to fix later.

How Does It Help?

Clean code requires a lot of effort, so why should you bother? It makes your future self thankful that you didn't leave the team with a steaming pile of garbage for a codebase. Your team members will also thank you when requirements change, and your code must be changed. The cognitive load required when refactoring can be drastically reduced if the previous person made their code easy to understand and change.

Let's look at some examples for context. In a clean code world, a function called fetchUserNameFromUserObject(userObj) wouldn't require further documentation because, as a consumer of that function, I would expect a UserName field to be returned from a given object. I wouldn't have to hunt around in the object or function to work out what to send; I just send the user object, and the username is returned. In an unclean world, the function signature might be userName(obj), which is extraordinarily unclear as to what it's doing. You might infer it's fetching a username, but it's very unclear. Is it going to return a value? What type of object do I send?

The goal of clean code is to convey the meaning or intention of your code or the work it performs as easily as possible. As a reader or a consumer of your code, you should never have to dig through code or read several comments that might be outdated anyway. You simply read the function name and the inputs, and it tells you what it does and what it should return.

The same goes for variables; naming them to something that conveys the right meaning helps the readers of your code tremendously.

Take, for example, the following:

```
const x = '1093'
const t = getData(x)
```

You won't know what this means or how to refactor it until you spend some time figuring it out.

```
const id = '1093'
const userInfo = getUserInfoById(id)
```

This is a contrived example. The first code is rarely written, although I've seen it. Without knowing anything about getUserInfoById, the second code allows you to conclude that given an ID, you will return user information. You don't care where the data comes from; you only care about getting the right data.

Clean code is about making it easier for you and others around you. You wouldn't write a novel using made-up words. Therefore, don't confuse your reader when writing code, either.

Resources That Will Help

There are several books I would recommend about clean code, which you can find at the back of the book in the 'Book Recommendations' section.

These are all invaluable resources that you should read cover to cover and many times over. I learn new things from these books every time I read them, and it's made me better each time. These authors are some of the best in the field, teaching you how they approach problems, write code and conduct themselves professionally. This last part is of great importance. Software engineering is so new that the field hasn't had time to adjust to the gigantic influx of people. As a result, there are some awful programmers out there. Your job is to conduct yourself professionally and produce clean, well-architected code that others can read many times.

You can learn a lot from books, but you learn the most from doing. Writing clean and maintainable code takes deliberate practice every time you sit down to do it. It won't simply be; it must be practised, refined and perfected.

CODE REVIEWS

Having your code reviewed

Code reviews are quite scary when you first submit a pull request (PR), and scary for good reason. I remember when I submitted one of my first PRs, I got over 100 comments from three experienced developers, tearing it apart. Why did you write it like this? You don't need that! That's just going to create more bugs. Now, they were all nice about it and assured me that it was nothing personal. However, I felt devastated, but after an hour, I had digested the feedback and realised it wasn't personal; they were just trying to improve my code and make me a better developer. Since then, I've gotten fewer comments on my PRs, but I still get them to this day; what developer doesn't? It's good to have people critique your code because they can see your code from a different perspective, and these alternate perspectives improve your code.

Much like a sculptor refining their masterpiece slowly over time, it takes iterations to improve your code, to make it more secure, run faster and look cleaner. That's why it's important to have code reviews. I know places that don't review code at all, and these developers struggle, and their skills never improve. Even though your first dozen reviews might seem brutal, you must take the comments on board, fix what you can, and ask for help on what you can't fix.

If being selfish and wanting to become the best developer you can be isn't a reason to have a code review, you might want to think about this. Code reviews allow you to check that the requirements are met for the problem you're trying to solve and that there will be fewer bugs, and one of the biggest ones is the cost of fixing defects. Fixing defects as early as possible in the life cycle of a software project is essential. An ugly bug that rears its head at the end of the project can be devastating, but the same one found early on can be fixed easily. There will likely be less code written on top of it, less code that relies on it, and more time to resolve the issue. Checking that you've met all the requirements for a ticket allows you to close it off properly. Having others inspect your code reduces the number of defects that happen. Code reviews are a cheap way to address issues before they become larger problems. They allow your senior developers to mentor the juniors, improving the team as a whole. As the saying goes, a chain is only as strong as its weakest link, and a team can't succeed if the weakest link is constantly introducing bugs.

Code reviews and testing are all part of improving the quality of your code; they are simple yet effective techniques. Code quality, consistency, and readability are all vitally important to the success of a project, and the less time you spend fixing bugs, the more time you have to produce more features and the happier the client is. The client expects zero bugs, and although that's impossible, you can go a long way in reducing the number you create. The more time you spend writing unit tests, integration tests, or whatever tests you can think of, the more time you save for your future self and others. It also allows you to see the problem solved, creating a discussion before it gets merged and forgotten for a while. It gives you a chance to discuss whether you have the right solution, if it's an issue, or if more difficulties arise.

Reviewing Other People's Code

I used to struggle to review other people's code. I figured they were more experienced or better, so surely I wouldn't find anything. Nothing was found, not because there was nothing to find, but because I wasn't looking. I wasn't trying to find anything; I would just tick and forget.

Reviewing code is hard. You must slow down, read it carefully, check that testing is being done, and suggest improvements.

Most importantly, reviewing code is a skill you develop. I am much better at reviewing code now than I used to be, but it took time and effort. I assumed you could just do it, but it turns out that, just like with everything, you need to practice. Once you have the skills, it's extremely helpful. You learn new things and more about the codebase in places you have never been before. As a lead, you are expected to make sure that the team is holding up the quality of the codebase. Furthermore, as a lead, you don't often write code. It's a better use of my time to review others' code than to write code myself. When you are a lead, you spend a lot of time in meetings, so it becomes hard to get the time to write code, but you need to be across the codebase. Reviews are where you do this. Every developer should review code; it's one of the requirements of being a developer, so let's discuss the process of performing a good code review.

The Main Things to Look Out for in a PR

Adherence to the Project's Standards

Every project should have standards. These are mostly enforced by a tool. Formatting and spacing should be automated, and each language has a tool to do this; some are built in while others are external, but either way, you need one.

Then there are standards in your codebase that aren't enforceable by any tool. These might be how you structure your files or write your tests. These should be well documented, and team members should be aware of them. I often find that people don't follow the conventions of the project not for any malicious intent but because they don't know they exist. I have had issues in the past where my team did not test well enough, and other teams working the same codebase complained. It was my fault for not reviewing the code properly. However, all it took was a 30-minute session where I went through how to test and what to test, and it was fine, and the team produced much better tests from then on out.

Testing and Testing Quality

Testing is hard, but it can be easy when you make the testing framework easier in your project. Most frequently, people don't add tests to a section of the application because it's hard, and it would mean figuring out how to test the core structure. Testing becomes easier if these problems are solved, and you're just adding incremental testing. I frequently find that the junior developers shy away from testing for this reason, and it makes perfect sense as they don't quite understand how you would set a certain section of the application up for testing. However, once it is there, it should be followed, this being another area of the PR to look out for. If a new file is added, it should be tested. If a file is edited to extend the functionality of a function, it should be tested. The more testing you do, the fewer bugs you will have. Speaking of bugs, if there is a bug and you fix it, add a test so that it doesn't happen again. You should look out for PRs that involve bug-fixing because a test should accompanies this bug-fix.

Code Quality

Others primarily read code; it's usually only written a handful of times but read many. This means all code should be easy to read and understand without too much mental effort or context. However, this isn't always the case. If I read code and the variables and functions don't make sense to me, I will ask for clarification. The author will adjust the code and make it clearer, making it easier for future developers. The reason you should focus so much on code quality is that it's something that decays over time. The more you leave the code quality, the worse it gets. You must constantly stay on top of it, monitor it, and improve it. If you let code be merged no matter what, the codebase will get worse. Over time, your velocity will incrementally get slower until you must rewrite the entire thing again! Quality is everyone's responsibility; thus, every PR you review should be high quality.

How to Comment

Commenting is an art form. If you are too harsh, you will put the other person on the defensive. If you are not direct enough, it will confuse the other person. It's all about asking questions in a non-threatening way. Ask why it was done this way, and if you have a suggestion, ask if they thought of doing it that way. Typically, they haven't thought of your solution, so they rewrite it, and it's much improved. Occasionally, they have a valid reason for how they wrote it, and it might be extremely complex, or the data structures are tricky. It might not be ideal, but you might be able to clean it up. It's about asking questions; never assume you have all the context because you don't. There are times when the code or functionality is done in a very specific way, which, when reading it with no context, is bizarre. When you ask the developer why it was done this way, they explain, and it makes perfect sense. Other times, you suggest an improvement, and they say it can't be done for certain reasons. Simply asking why is enough to get the other person to think about what they are doing. It could well be that they never thought of doing something a certain way, and they thank you. This allows them to grow. This is what commenting on a PR is all about, allowing the other person to grow and improving the quality of the codebase. It should never be done in a harsh or personal way. You should always have a curious mind and comment with empathy.

Don't Get Caught up on the Little Things

Everyone has their quirks regarding development, but don't get caught up on the small things. You don't need to pick apart their PR; you just need to go through it with an eye for quality. Anything minor should be left if you feel it's a deviation from the project and could become a habit. Raise it subtly and politely and explain the situation and move on. However, don't tell them to remove a space here or a line there; this is a waste of time. Your goal is to improve the code, catch anything they might have missed, and move on. Anything big should be picked up and either addressed or noted. Anything small should be ignored.

TESTING

Testing is one of those things that, as a graduate, you don't put a tonne of thought into. You've had lectures, and senior developers tell you it's important, but do you listen? No, of course not; you know better. Besides, you can't create new functionality if you're always testing previous work. You can't move onto the next card if you're stuck testing the old one. You don't feel productive when you're not adding new features or fixing bugs. Yet, the longer you go without testing, the worse it becomes. I went into my first job thinking I should test every line of code, just like my lectures and books had taught me. Yet, after a few months of doing some terrible tests that didn't test much, I gave up on testing and told myself I'd get to it later. Of course, you never get to it later, adding twice as much of your time in the long term when you have to revisit it because it doesn't work as it should.

You must learn things the hard way. I can tell you all about how I didn't write tests, and it came back to bite me, but until you go through the process of failing to write tests, you won't learn. You just have to learn for yourself, and you will. Your first year is one big learning experience. So, only when you spend time around better and more experienced developers do you realise that you don't know anything. Testing is one of those areas where you find out how important it is and how much of an art form it is. Not only do you find out its importance, but you also learn about the complexities and nuances of creating good tests.

Test Types

Clean code is never void of tests. You should have extensive levels of tests across a wide array of your code. Typically, this will be in the form of unit tests. As servers have become more powerful for cheaper, integration and end-to-end testing have also begun to play a bigger part. In today's age, integration level and end-to-end level testing are just as essential as unit testing, if not more important. You should have high-quality tests that span all your codebase, and the ratio of the unit to higher level tests should be skewed more into the integration and end-to-end realm. In today's age, we have huge moving pieces with many connections or dependencies on other systems. Gone are the days when you just write a small program that sits on your computer and does something. Now you must connect to the cloud, payments, email, etc. There are magnitudes more moving parts, so integration tests to capture the errors between these moving parts are crucial.

Static tests: these are not tests but more part of your code. These could be languages with strong type systems. A type system would verify that the type of data going into one function matches; otherwise, it gives an error. These tests are typically done inside your IDE or the linter you use. Behind the scenes, you have some software that compiles and runs traces through your code to ensure you pass an integer to the calculateTaxForUser function rather than passing a string that would blow up. These are not tests you write but come with the tool you use, think type systems. You won't set out to run the static tests; they get checked behind the scenes. You can also run tests in an automated fashion in a pipeline. This would include that the code is formatted correctly, the types match, switch statements have a default condition, etc. These are all static tests, which can catch bugs before running your code.

Unit tests: these are small tests on a per-function basis. I test the inputs and outputs of each function I write. Given the right conditions, I check that the conditionals do what they should. Furthermore, I check the correctness of each function I write. Unit tests are primarily used to confirm the business logic of a system. If I had an accounting system, it would check that the function calculateTaxForUser calculates the correct tax for a user. Unit tests are vital for identifying the bugs that lurk in these functions. These tests are simple to write and maintain and effortless to run.

Integration tests: these are larger tests that string together multiple parts of your system. This could be hitting your API from an HTTP level down to your database layer, checking that users are input into the database correctly when they sign up (at an API level). They can also check that when I hit the sendEmail endpoint, an email or confirmation is sent back from the API to signify that an email has been sent. Typically, integration tests are as real as you can get them to give you the best results. You could also have integration tests that run off mocked data when you rely on unreliable external systems. I have used this in the past, and it works well, but it will never be as good as the real thing, but that isn't always possible.

End-to-end tests: these are the largest of all the tests. In the web development world, these tests would run automated tests through your front-end to your back-end and check that the front-end gets the right results and behaves correctly. End-to-end testing is something that has only become easy enough to use for all projects in the last few years. There are fantastic tools out there that allow you to mimic a user completely. You can capture every possible action they could take and run it in an automated way. These tests are extraordinarily valuable, but they can also result in the most flakiness, which is a term that describes how unreliable a test is. The idea is that you don't have a lot of end-to-end tests, only enough to cover the most used paths in your system. Then you leave the other paths to your other testing levels. Even with great testing tools, end-to-end testing is still difficult because you are so high up on the system; you're at the user level, so if anything goes wrong, you see it. Therefore, tracking the most used paths is great to ensure that all your error cases are properly handled so that a user doesn't see a big wall of red text when something goes wrong.

What Percentage of Tests Should I Have for Each Level?

I believe that you should have a lot of integration-level tests. You should have enough unit tests to test the overall correctness of your business logic and end-to-end tests to pass through the major user flows. However, the real complexity in most systems is the integrations between systems, which is where your integration level tests shine the most. These should make up the bulk of your testing suite. They should be easy to run and maintain and run in an automated fashion. This is often referred to as the testing trophy. The more integrations into your system and others, the more you need to test they work. If there is a bug in the function to calculate tax, it can be easily picked up and corrected if the logic is in how the tax is calculated. For example, say it used the wrong tax brackets; you could easily figure that out and write a test for it since you are getting the tax information from another system. However, if you don't have control over that system, you would have to raise the bug with them. If you have control of that system, you would change that code, write tests, and write more tests in your system to ensure it won't happen again. The effort to fix a function calculating something incorrectly over another system sending the wrong data is magnitudes different. Therefore, having an integration test that can verify that this other system is sending the right tax information is critical. It allows the team to be confident that if the tax system pushes changes to their code if it breaks anything on their end, they know about it quickly. This leads to the next point about testing.

Automating Your Tests

There is nothing worse than writing tests, only to find out that you had never run them, and a week later, they are out of date. Tests should always be run automatically in a pipeline that is triggered whenever someone checks in new code. We live in an age where pipelines are cheap, easy to use and run without so much as a click of a button. All you need to do is check your code in, and a suite of tests will run. This verifies that the code added won't break the rest of the system. Not having automated tests is a modern-day sin; you must have a system that regularly runs your tests, one you don't have to lift a finger to run; it sits there waiting for someone to check in their code. This stops thousands of bugs, your tests are never outdated, and you instantly know when something is broken. Having an automated system is a lifesaver, and it's worth every penny.

However, automating your tests is not easy. Yes, the tools themselves are easy to use, but creating pipelines to fit your complex needs is hard. Let's take unit tests, for example; they are easy to write and run and typically do so with one small command. You don't need to spin up other services, make sure a database exists, or verify that a third party is doing what it should be doing. You are testing at the per-function level, meaning they run extremely fast and are straightforward to automate. This allows you to set up your pipeline and add the command, and you're good to go. The real complexity is running your integration and end-to-end tests, which might require running other codebases or fulfilling certain conditions. This is where the fun begins. It takes a lot of time to create, but it is more than worth it. When it comes time to check in some code, you can see hundreds of green ticks verifying that the most common user paths are good to go! This gives you a level of stability that you just don't get from unit tests.

Early in your career, you won't need to stress much about these as they should be set up by others or already in place. You should understand how they work and how you should be thinking about them. One day, you will have to build one, and you will need to understand how they have worked in the past.

CONTINUOUS INTEGRATION AND CONTINUOUS DEPLOYMENT (CI/CD)

Continuous Integration and Continuous Deployment (CI/CD) is the technique of checking in your code into the wider solution at regular, rapid intervals and pushing to production as rapidly as possible. CI/CD is the ultimate in agile software development. It was once the case that you wrote code for months, and then right before you finished all your features, you would push it to an environment for testing. Today, computers have become cheap and tools effective enough that pushing your code into a hosted environment is simple, cheap, and catches many bugs. Furthermore, the CI part of the equation is about getting your code checked in and tested with the rest of the codebase as often as possible. This happens through automated testing and code reviews.

Continuous Integration in Depth

CI is critical to any software team, and the goal is to merge your code into the rest of the codebase as frequently as possible. Gone are the days when you spend weeks on a feature, merge in other people's code, and it all breaks. Today, you should be merging your code in with a goal of daily intervals. This reduces the chance of conflict, issues and diverging from the codebase. This is why we have agile development; each two-week sprint contains tickets, and each ticket should be broken down into daily or fewer tasks. The smaller you can break down these tickets, the more often you can check them in and the fewer chances you have with code breakages.

Branching

Traditionally, with modern Git-based source control, you would follow branching. This is where you create a branch that breaks off from the main branch. In this branch, you perform the work required to complete your ticket. Once you have tested this, you put up a pull request for others to review. Once this is reviewed, you merge it into the main branch, and you are done. This is the typical life of a ticket. Now, where it gets complicated is when you have many branches. You might have a branch that tracks a development environment, one for staging and one for production. You would then merge your PR to develop, staging, and production. This is a way to keep your code out of production and safe. This works quite well but can often become slow and cumbersome, and when the team grows, it gets increasingly slower to release. Ideally, you want your work checked out by others and tests passed to be in production. It doesn't have to be active for a real user, but it should be in production. This is where trunk-based development comes in with continuous deployments.

Trunk-based Development

Trunk-based development has gained popularity recently with the rise of CI/CD. In an ideal world, when your code is approved, tested and done, it should be in production as quickly as possible. Why wait days, weeks, or months for it to get there? In trunk-based development, you typically have one main branch. This is where all your code is merged into via PRs, or if you are a company with pair programs, you might skip the PRs altogether and check it in when both parties test it. Once it is in main, it's up to your pipelines to automatically test your code, and if it checks all the right boxes and passes everything, it is pushed to the first environment. When this is also checked automatically, it is pushed to each environment until it reaches the production environment. Ultimately, it depends on how many environments you have, but the goal is still the same. Once it has been checked into the main branch, it should be in production once it passes all the right checks. It should involve zero human interaction. If your code is a feature that is half complete for whatever reason, feature flags come into play. A feature flag is a code; think of it as a yes or no boolean value that wraps a bit of code to turn it on or off. Feature flags are fundamental in CI/CD and trunk-based development. They are fundamental because they allow you to push your code into production, half-baked, not fully working, but in a way that won't affect the rest of the system or the end user. Your code is off by default but always sitting with the wider codebase. This leads to less time worrying about merge conflicts and more time getting your code into production. You have the safety net that it's not on from the feature flag. Then, when it comes time to turn it on, you simply flick the switch, and your code is live. No separate deployments or releases are needed; just a flick of a switch, and it's on.

PLANNING THINGS BEFORE YOU START

Planning a Single Ticket

One thing I continually try to do before I start any ticket is to plan what must be done. Typically, I open a note on my laptop, list the requirements and paste the ticket link. I'll then write a description of the ticket from my perspective and understanding. Then I think about all the steps I need to take to get it from start to finish. This includes the tests and code stages I hit, any packages I need, or people I might need to talk to. Once I have captured all this information in the ticket, I look through the codebase and find where I must put the code and whether any functionality matches this code already. This is key because, too often, I see people skipping this and end up implementing something that already exists in the codebase, or there is a common package to do this. It's important to reuse the code that already exists in the world as much as possible, as this cuts down on bugs and time. I find that this planning time helps establish the scope of the ticket. The larger the ticket, the longer I spend planning the ticket. Once I have spent time planning it, I can easily finish it because I've thought out how I will write and test it and what the requirements are in advance. This also allows me to see if there are any major pitfalls in the ticket. Perhaps the ticket didn't account for an error scenario, so now I know to ask the person who wrote the ticket what I should be doing in this situation. This is very helpful **before** you start coding because it allows you to capture as many error cases upfront. The more you plan, the easier the work and the

147

more obvious the pitfalls. There have been times when I would plan a ticket and look at the estimated time to complete it, but through my planning, I realised the ticket was impossible or would take significantly more time than expected. This can often happen, but the best part is that you can raise it early and not after spending multiple days on it. It helps to give your boss more visibility on the ticket and explain why you expect this to take significantly longer than planned. Your boss will love you for this; I promise you!

Planning a Larger Feature

Before we start any major piece of work apart from a minimal feature, we must sit down and plan what we need to achieve and how we will do it. The act of planning a feature is extremely beneficial. It allows you to go through the process of planning the requirements, looking at the current implementation, if there is one, researching the trade-offs of each approach, and gives you a chance to show other people. Showing others what you're thinking before you do it is more important than one might think. Showing someone else gives you that sanity check; it allows someone else to come in, understand what you're doing and point out things you might have missed. This is okay because you haven't written any code, so it becomes straightforward to start again.

I once had to implement a payment system using Stripe, which was much harder than I originally anticipated, and I knew it wouldn't be straightforward when I started. We use Confluence to create wiki pages to plan our features. Not only do they act as a plan for others to see, but they also act as documentation beyond what is in the tests and code. I spent about a week planning, weighing the various payment vendors, how we would roll it out, what the requirements were, and whether we needed all the features in for day one. Then, once I finished, I could show the whole team, and they could read over it, discuss it with themselves and most importantly, raise any issues they had. Beyond the issues, they also raised additional problems I had not considered or better ways to sort them. This collaboration was fantastic; it meant that I covered all the bases and thought of all possible difficulties. Most importantly, the feature was documented so that people in the future could know the thinking behind why we did what we did and how it worked.

This documentation lives in a hosted central place where anyone in the team can log in to find the plan for the payment feature. Furthermore, it makes it much easier for people in the future to make changes because they know how and why it works that way. They can look at the features of that system and add new ones by following the process already laid out. One of the benefits of planning is that when it comes time for someone to review your pull request, they have additional material to explain the thought process behind your design decisions. It makes it easier to check the code's features to the plan's features, ensuring they're all there and working as planned. How often do you review someone's pull request only to become lost trying to understand what they have done? You might still get lost, but at least you have some plain English to look over before you jump into someone else's code.

Planning takes time. You sit there, think about and research the problem and never write any lines of code! This feels unproductive and meaningless, but it's not. When planning, you think of all the pitfalls, the edge case to account for, and create a higher quality outcome as a result. I would argue that it's a much better way to spend your time than jumping straight into code. It forces you to slow down, think about the problem, and find gaps in the feature requirements and description. You're forced to think about why you're doing this in the first place, beyond being assigned to it. It gives you time to think about why this feature benefits your customer and how to make it as easy as possible to roll over with as little downtime as possible. It allows you to ask questions, which is important, especially when someone with zero experience with development has written the features. They could be asking for something that isn't possible, but you might not know that until you get deep into the code or spend a day planning it. You find out that the feature they want can't be built into the system because that's not how the system works. This saves you time in the future.

Lastly, planning what you're going to do produces much less error-prone code. It's simple: the longer you spend thinking about the problem, the trade-offs, and the inevitable limitations, and you discuss it with others, the better your code and your product. You also spend much less time fixing bugs or adding additions to the feature because you left something out.

I've spent many days creating a feature only to push it into main and find out that I've missed some key functionality that the front-end needs. However, no one realised because I never showed a front-end developer the plan; there wasn't one. Conversely, when I had a plan, the quality of the feature and the smoothness of the rollover have been much easier even though it might have taken longer.

Producing high-quality code is vital. It creates a better product, fewer bugs and easy-to-maintain code. All these things are possible without a plan, but you will have a hard time doing them. The more you plan, the easier time you have when it comes to implementation and maintainability. You spend less time fixing your terrible code and more time looking like a brilliant programmer.

HOW TO LEARN EFFECTIVELY

Learning Needs to Be Fun

In school, you spend 15 or so years learning. Each day, you were filled with knowledge and tested on it, but how much did you retain? Why can't we remember anything we learned in school? Because it was boring.

I came to learn the simple fact of having fun while learning when I was attempting to learn French. The language sounded sweet to the ears but also like a challenge. I spent some time learning it myself, and it was fun, but it slowly became less fun. I hired a tutor, and he taught me not to force it. I wouldn't learn nearly as effectively if I did not enjoy it. Thus, I began to entertain myself with my learning; I would mix things up with all sorts of things, from exercises to movies to YouTube. You name it, and I would try it, but I wouldn't stick to anything when it got boring. I learnt a great deal of French, and for the most part, I enjoyed learning it. I take this same technique and apply it to self-study in the field of software engineering. I never spend my leisure time learning algorithms or some piece of technology unless I'm interested, and on the flip side, I can spend hours googling and reading about whatever interests me that day. This sort of learning isn't done to get to any place or to pass a test; it's simply to learn for learning's sake. This is the objective because you never know when something will be useful.

Use Your Interests to Make It Easy

I was once asked what I do to learn things, and when I replied that I google things, the person asking looked at me like I was an alien.

Most people are set on school-style learning, where taking courses and getting a diploma is all that counts. This works for some, but it didn't work for me. The structure of these courses is okay, but they can often be dull or cover more content than you care about. In this sense, I find structured courses useless, whereas I find googling and working out problems much more rewarding.

Most of your in-depth learning is on the job when you are given some technology for a project and are forced to use it daily. If you wanted to quickly gain in-depth knowledge about this topic, you could do a course or work your day job. However, when it comes to self-learning, you should only learn the topics you want to learn and learn in the way you feel is the most enjoyable. If you enjoy watching a YouTube tutorial for five hours, then certainly do it. If you like taking short weekend courses, certainly do it. Find the most enjoyable learning method and use that to your advantage. The more you learn, the more you command more money and responsibility. The more you build the habit of following your interests, the more you learn and enjoy learning, even if it isn't remotely related to your work or career path. Don't just cultivate better skills in software engineering; instead, research how to run teams or deliver software faster and more reliably. Figure out what makes the best teams tick and replicate it in your teams.

Learn a Little Each Day Using the 1% Rule

When I was a teenager, I read many self-help books because I was obsessed with getting the most out of myself and earning the most money. Once you have read five to ten, you've more or less read them all. However, the best I read was *Atomic Habits* by James Clear. It talked about seeing your life as a set of days; to get the most improvement over the long run, you need to get 1% better each day. As long as you keep this in mind, you improve at a compounding rate. One per cent better than the day before isn't much; maybe it's a few articles, adding a new tool to your belt, or spending some time getting better at public speaking. One per cent each day over an entire career of 40 years is incompressible. Bill Gates has a famous quote:

> 'Most people overestimate what they can do in one year and underestimate what they can do in 10 years.'

This quote is awe-inspiring because we forget how far we have come in 10 years; we only look at where we were last week. Each time I got a raise, I would compare it to my current salary and be disappointed or mildly happy at best. I was never ecstatic with the new salary or position. It wasn't until I began looking at how far I had come, existing in a state of gratitude for my position, that I began to feel happier. In my first five years, I went from a starting salary of $60k (AUD) to $290k (AUD). If I looked at each raise I received, it was substantial, but it wasn't mind-blowing. However, the moment I looked back on what I was earning and then looked one year further back to the time I was making $15 an hour at the local Subway, it was mind-blowing.

This leads back to the 1% rule; making incremental improvements each day eventually leads to considerable rewards.

When I started, I was a back-end developer in the webspace, using a fantastic language, but ultimately, it was quite niche. At the time, I was limited in the types of projects I would be assigned. This meant that to be more valuable to the company and ultimately command a higher paycheck and title increase, I would have to know the work that was in demand, and at the time, it was front-end development. I taught myself numerous front-end technologies and more useful and widely used language in the back-end. As a result, I could go to more clients and move companies for a salary increase in that newly acquired language. It took around six months before I was confident enough to get a job in this technology, but I got it, and ultimately, it was from my constant self-study because I was interested in it. I have used this technique my entire career, constantly looking for things that interest me or technologies I can add to my toolbox. The desire to pursue whatever interests me has been one of the biggest factors in my salary increases. You must remember that 90% of developers won't put in a single minute of extra learning or work outside the 9-5, so putting in even 30 minutes each day compounds, putting you above the rest. You also find that your career skyrockets, and you can confidently go from job to job, interview to interview and ace them. That is the other thing about interviews; the more you know, the easier they become. You can ace interviews because you will be asked how you have done this or that in the past, and you may not have done that in a workplace. However, you can say you haven't done this in a job but have done something similar in a side project and talk about what you did, which will more than satisfy them. You never have all the answers, but at least you can get 80% of the answers via something you did in the workplace or your time. I have answered many questions like this, and it always goes well; they're always impressed if you spend time on a side project.

Now, I want you to think to yourself, what part of my skill set could I improve BY 1% each day to see drastic improvements over time? Then think of a way you could improve this, think about the technique of learning that best fits you, and then go out and do it. For example, learn front-end when you have only done back-end development. If you find learning from YouTube videos in a structured tutorial-based approach easy, find a series of videos that teach the fundamentals of front-end development. Follow along, code along, have fun and if you stop enjoying it, find another method or choose the topics you are interested in. Nobody says you must do every topic or redo the topics you might already know; just skip them and continue. Most importantly, you are learning something new that makes you better each day.

ALWAYS LEARN NEW THINGS

Read Everything and Anything

I am always reading something, whether it be finance, something spiritual, or fiction; it doesn't have to be software related. However, I always have a book on the go. The skill of reading and pulling out the most relevant information is something that will serve you well throughout your career. It's also fun and an easy way to wind down from the day.

Compiling a list of books in your field is the best way to always have something on the go. You can learn a lot about abstract software concepts or how to design systems, run teams, manage individuals and get better at writing code, all from reading books. It's a low-effort activity that can get some knowledge in your head to recall something later. Being exposed to new topics and reading a little about them causes you to think differently and stretches your brain to solve problems from an alternate perspective.

You don't just need to read books; you can read random articles online or subscribe to a weekly newsletter. Just search for topics of interest and read about them, and read for the enjoyment of learning.

Build Weird Things

I spent a weekend loading custom firmware onto my router and then loading ad-blocking software onto it to have ad-blocking on my entire network. It didn't take long because most of the work was done by the people who wrote the software; it took the most time to SSH into the router. I had to brush up on SSH, dig through my router's instructions and sort out the terminal. These were all things I had done before but partially forgotten. I had a blast, and I most likely didn't learn anything that would help me in my day job, but I learnt a little more about networking and routers, which may or may not be useful in the future.

I knew someone who built a garage door opener with Go. It took him a few weekends, but he got it done and could open his garage door with his phone whenever he pleased. My dad is notorious for doing this; he is an electrical engineer by trade and has a background in computer science. He is also fascinated with woodworking, so he often combines it all together. He once made a wooden clock with an e-ink display for the time, all programmed via an Arduino. My dad doesn't do these projects because he needs to for his job; he does this because he has a passion for making things. Your passion shouldn't be for programming or a framework but for creating, making, and building. These passions last an entire career, and you can think back on when times are tough and take pride in what you've built.

Watch Videos

YouTube is one of the biggest time wasters; however, if you use it for good, you can learn a lot. I frequently use it to find topics I'm interested in and then dive deep into them. I don't just do it for tech; I do it for everything! No matter what I am keen to learn, I go to YouTube, and there will be a video on it. From replacing a vacuum trigger to fixing something in your car, YouTube has just about everything. There are many fantastic channels to provide you with daily information to keep up to date or entire channels dedicated to teaching topics. I'm not a handyman; in fact, I am useless, but for some reason, I decided to paint my house or at least a room or three. I spent hours researching painting and watching painting videos, and in the end, it paid off to the point that I received many compliments from painters on the quality of my work. All this from watching a couple of hours of videos on YouTube and talking to my father and father-in-law.

Learn from Others

One of the best ways to learn anything is to talk to someone with experience. I have talked about this in my mentor chapter, but working with someone already incredibly knowledgeable about a specific topic pays off. By doing so, you will get better. So, soak up as much information from those around you. The more you can obtain, the faster you will grow. The faster you grow, the more opportunities you will get to earn more money and work on great projects.

FINISH WHAT YOU START

We often spend all our time in the code, spending hours trying to debug problems until it drives us mad. We finally get it done, and then it's time to test, clean up and PR. It's tempting to push your code, raise a PR and call it a day. However, this will not fly at a good company because you always need to write tests, clean code and improve something in the codebase. This is hard to do and stick with, but it doesn't take much to write the tests; it requires discipline. You have to slow down, take a breath and realise that the extra time you spend cleaning up your code, writing tests and fixing some messy part of the system pays dividends in the future. The more time you spend improving what you have in the present, the more you benefit in the future with easy-to-maintain code.

Many developers skip this step, where they aren't productive unless they put tickets into the done column. This is not what software development is about; instead, it's about building great quality software and not caving to the time pressures to make it. Deadlines are always tight, and you might save some time in the short term by skipping these sorts of things. However, it quickly comes back to bite you when you continuously ignore the acts of a professional. It might take a few months, but before long, you will notice that your productivity slows down and you begin to take longer on tickets. More bugs pop up, and more time is spent fixing these bugs with further band-aid fixes. Soon enough, you're in a situation where the code is unmanageable, and you can't possibly reason with it. This is a dangerous place to be because this is where people start to talk about performing an entire rewrite of the system. This is not where you want to be, nor where your managers want you to be. You can avoid this, and it's simple: you write tests and spend time making sure your code is well written, easy to understand and follows the conventions of the project. If you take 25% longer, even 50% longer, on tickets, but it results in a solution that actively improves the codebase rather than makes it worse, then that's worth every minute of it. You can't spend eternity on tickets, so you need to figure out what is worth fixing now and what is not. You need to look at the time to fix it and the cost to fix it later. Is it a function that has a wrong comment? Fix it now; it takes three seconds. Does a major part of the system have to be rewritten because requirements have changed? Don't fix it now; add a ticket and prioritise it. If you fail to prioritise it, you won't improve anything. The sooner you resolve these issues, the better, and this is a case of something that will come back to bite you if you keep putting in hack fixes. You have limited time and can't fix everything, but you can make small changes that slowly improve the solution.

The Boy Scout Approach

In Boy Scouts, they teach you to leave a camp better than the way you found it, and this can be as small as putting rubbish in the bin. Over time, as people continue to do this, the campgrounds look extraordinary and take little to improve. This is the same with your codebase; the more incremental improvements you can make, the better your codebase becomes. More importantly, the easier and more productive your life will be. When you add a new feature, there might be a function that you use to generate the data for your front-end component. When working with it, you notice a bug, and it is quite hard to understand what it is doing. Fix the bug and tidy up the function. It probably won't take you longer than 30 minutes to an hour, but your future teammates will thank you. It's these small improvements that add up over time. I have seen a codebase that was awful to work with, complex, messy, and hard to change and over time, with mostly small incremental changes, it turned into a beautiful codebase that you could hang your hat on. Now, don't get me wrong, there are still going to be times when you have to make drastic changes that can't be done in small increments. Depending on the age of the codebase, there might be many, so you need to set aside dedicated time to do this. Stop putting it off and advocate for your codebase to be improved. The more you advocate, the more people will listen, and eventually, people act, and the codebase is improved. Don't settle for the codebase you have, the complaints you have with it or the slowness of development. Take it into your hands and improve it without remorse. Your future self will thank you.

Rewrite

A codebase rewrite is a scapegoat for poor management and leadership. However, not always; some codebases are old and use an outdated language, so porting it to the latest version would be more work than just building it again. These rewrites are okay; they make sense. A codebase can't exist forever if you intend to keep changing it. Conversely, I have seen companies rewrite their applications multiple times before going to market. This is because of terrible leadership, unprofessional programming and a lack of knowledge. It's not the developers' fault; it's always the person in charge, whether the tech lead or the managers above.

A rewrite is the opposite of a professional but is, at the same time, frequently required. You can't control how a codebase was built four years before you got there, but you can control it now. A rewrite should be the last resort, as you can salvage most codebases with enough time and effort. You never have a perfect codebase, but you can certainly improve it. You might not be able to do so within six or 12 months, but you could in 24 months if it's gigantic. Improving a codebase comes down to a mindset shift. You don't attend work to finish a ticket and leave.

You rock up, complete the ticket to the best of your ability, and resolve issues along the way. If everyone picks up a few bits of rubbish, the campgrounds are clean in no time, and most of all, they stay clean. Continuous improvements are a habit that is encouraged from the top down.

DON'T RUSH

The worst code I ever wrote was when I was rushing. As a junior, I did many stupid things; I wrote horrible code and didn't test database migrations properly, which ultimately broke a production database, produced security holes, etc. I distinctly remember when I created a migration, and it worked fine locally, but I didn't account for the change in the names of some fields. Elixir and its spectacular library, Ecto, can handle a lot of the migration and rollbacks for you. However, months prior, I had named two fields incorrectly, like fields-one and fields-two. It should have been field-one and field-two. Now, the migration changes them without a problem, but Ecto can automatically roll back many changes; it's smart in that way, but it's not God. It can't roll back everything, especially when you change the name of a field. Of course, locally, I would migrate but never test the rollback. I also didn't try the migration on the production dataset. Why? I was rushing, of course. In production, it did the migration, failed, and then the rollback failed. So, I called the senior developer, freaking out, of course, as production was now completely down. We spent the next few hours fixing the production database and getting it back in shape. The hours wasted could have been avoided if I hadn't rushed. However, as a junior, my goal was to produce as much code and complete as many tickets as possible daily. I wanted to prove I was doing a good job. This, of course, resulted in me rushing through tickets and code and not testing properly. In the end, I learned how to fix the issue and learnt a valuable lesson about not rushing.

It doesn't matter what level you are; you need to take time to complete your work properly. It doesn't need to be perfect because it won't. Furthermore, it does need to be properly tested, thought out, planned and peer-reviewed. The need to rush out my code was entirely created by myself. No one ever told me I was too slow. However, when I stopped rushing, I produced much better code and began to gain a lot more respect.

Programming with patience is key. Rushing through your work only leads to trouble.

BE HUMBLE

The hallmark of a great leader is someone who can extract a greater output than the sum of a team's parts. Tim Duncan of the San Antonio Spurs is the hallmark of a humble leader. He was the pillar for the most successful team from the late 90s to the early 2010s in the National Basketball Association (NBA). The Spurs won the championship five times with Duncan at the helm, yet not once would you hear him take the credit for winning despite him being one of the greatest players of all time. His humble, team-first nature permeated throughout the organisation, giving them the reputation of having the most beautiful offences in the league. His quiet, calm leadership kept them cohesive for 20 years. He was also instrumental in mentoring younger players, as was the entire organisation, which extended the time the organisation was dominant.

'I'm surrounded by nothing but great people. I've been blessed with that, so really, I've got no choice but to be an all-around good person.'—Tim Duncan.

This shows that even though he's one of the greatest players to ever play the game, he attributes it to everyone around him. No matter how many NBA championships or individual awards he won, he still attributed everything to everyone else. Tim Duncan was not only one of the greatest to ever play the game, but he was the most universally adored. Finding someone with a negative opinion of the man would be impossible. The only negative thing you could say about him was that he was too good.

I resonate with humble leaders—those who are first to bring other people up and give credit. As a junior developer, when you work with someone more senior than you, and they are first to give you credit is motivating, reinforcing the notion that the work you did was noticed. On the other hand, if a senior developer takes all the credit and doesn't praise anyone else, then you would avoid working with them again. The same happens in basketball. When one player takes all the credit, the entire team disengages. Team chemistry can't be destroyed any quicker than when one person takes all the credit. The longer you take all the credit, the more people won't want to work with you. You might get recruited for big roles in higher positions, but you begin to see less success as you alienate the people around you.

No matter how good you are in whatever you do, you won't be perfect or the best in your team for everything. There are people better at something, and if you can't step aside and let them take the credit, you won't have a team much longer. The thing about giving credit to others is that the more credit you give others, the more you get back in return. You might be on a team and worry that people don't know what you did, but if you're on a decent team, they will compliment you as much as you compliment them. Being humble propagates into the group as a whole; it brings the team together. It's about the good of the team and not the good of the individual, so the more you humble yourself and put the team first, the more success you see from a team and individual perspective. You don't have to be on one team for long because others will want you on their team before you know it. The more you do for the team, the more the team does for you. The more you humble yourself, the more you become someone who can positively contribute to a team, life and those around you.

Life isn't all about being the most successful person around. Life is about being the best person you can be, and the humbler you are, the more people want to be around you. You will struggle to have meaningful relationships if you constantly tell people how good you are or take all the credit for team achievements. On the contrary, if you praise others first and rarely discuss yourself, people will flock to you. People want to be around others who make them feel good. You can feel it quickly when someone thinks they're better than you or the people around them. They might produce incredible work on an individual level, but people become put off by them. People won't work extra hours or go the extra mile for someone who's arrogant and thinks they're superior to everyone. People go the extra mile when they respect and admire someone, not when they're told to.

Quiet Confidence

Being confident is an absolute must in any knowledge industry. There is no room for self-doubt, pity, or a can't-do attitude. People can sense this and immediately look at you differently. If you struggle with confidence as I do, you must become inwardly confident to become outwardly brilliant. You must practice confidence because you can do it; you just have to set your mind to it and believe in yourself. You must become quietly confident, meaning you can tackle anything you set your mind to. Someone who is quietly confident can be given a task, and they may not know what needs to be done, but they are so confident in their abilities and knowledge that they get it done. When you are quietly confident, you never feel the urge to blurt out how good you are at something; instead, you let all your work do the talking. When you are quietly confident, you get the respect you deserve without demanding it. You are seen as someone who can tackle any task and produce great results.

Maintaining a balance between being confident and staying humble is not easy. It involves work to build your confidence and remain humble unless you are that way naturally.

Reflecting on Your Achievements

You should make it a habit to regularly reflect on your achievements. This could be done by updating your resume, looking at what you have done and thinking about these experiences. You could also set some time aside each week to reflect on what you have done in your life. I mention this because this is how you develop your quiet confidence. You cultivate a list of achievements you are proud of, take them and ruminate on them, and this drills into your self-confidence. As you perform this regularly, your brain slowly becomes more confident without you thinking about being confident. You don't become confident from saying you are confident; you become confident from achieving and putting in the deliberate practice to get to the level of someone that can achieve these things. The more you practice, the better you become, and the better you become, the more confidence you have. It is a slow process performed over the years until one day it clicks, and you are what you always saw in your senior developers—someone who has most of the answers, can solve problems quickly and produces great quality code. You won't have to tell people about this new-found realisation; they have known for years. You can be great; you just have to put in the deliberate work to get there.

This is something you must practice, as it doesn't come easily. The more confidence you gain, the more you want to tell people how good you are. You must resist this urge at all costs, as it will undermine all your work. You must remain quiet, confident, and relaxed and continue producing great work. Eventually, you will be that great developer, and people will know; you won't have to tell them.

You're now at the stage where you have a reputation for being great, can tackle any problem, and output high-quality work all the time. You must remain humble, no matter how much praise you receive. This much attention can be hard to deal with, but it comes up as you rise through the ranks, especially when you get direct reports asking for help. Being humble gets you more respect, you have better team cohesion, and you work better with others. On the contrary, if you are arrogant, overly confident, and constantly tell people how good you are, it has the opposite effect. People don't like arrogant people; they're not fun to be around.

However, you might be wondering if you've been overlooked in the past because you weren't the outspoken, confident type. Yes, because you didn't have the confidence, to begin with. Those who are arrogant or confident are either faking it or have it but express it poorly. Either way, they attempt to gain the trust of others by appearing confident. In the case of the arrogant type, this works in some situations; in the case of the confident type, this works fantastically. The confident type is someone you want to be. People can sense the passive natures of others, so when someone sits back and lets others take charge, they garner less respect. When someone takes charge, they get respect because they appear to have better skills. A confident nature can take you far, so you must cultivate an aurora of confidence. As mentioned, this can only be achieved one way: practice. The more you practice, the better you become, and the more you look back at how far you have come, the more you realise how good you are. The more you realise how good you are, the more confidence you have and the more respect you gain.

TEAMWORK

Working in an effective team is the same as having a harmonious and successful marriage; you must work out your problems and communicate with each other, and you will fight.

I've worked in teams for most of my life, having played basketball since I was eight or so. Team sports have strengthened my skills in working as a team and working with others to achieve a goal. No matter what happens, you always have your team by your side, which holds true in the professional world. Unless you're working on a project by yourself, chances are you will be working in a team. You must get used to other people. You don't have to love them or even have to like them, but you have to respect them and work effectively with them. You can't do everything yourself, nor would you want to; it's a lot of work. Instead, you will have to delegate, work together, lean on each other's strengths, own up to your weaknesses, and not kill each other. This is all teamwork, so the better you get at teamwork, the better your career becomes.

Understanding Everyone's Strengths and Weaknesses

Everyone has strengths and weaknesses. Understanding how you benefit the team and how others benefit the team makes the team more cohesive. The chain is only as strong as its weakest link. When you have someone who works exclusively with JavaScript on the front-end but gets put on strict SQL, they won't be highly effective. Having developers work on their strengths leads to better results than having them work on their weaknesses. Leaning on the strengths of others helps more than just you; it helps the team as a whole. It saves time, money and more time. If you believe your strengths lie with other tasks, insist on working on those tasks. If you have noticed that a developer is unhappy with one type of work, move them or reach out to someone who could move them. Teamwork isn't just about effectively working together; it's about knowing what roles need to be filled and then putting people in the opportunities to succeed. A fish feels like a failure when it's told to climb a tree.

Rules of Being an Effective Teammate

Rule One: Everyone Has Something to Teach

It doesn't matter if you're a newly graduated computer science student or have 40 years of experience as a tech lead; you have something to learn from everyone. The sooner you realise that everyone can teach you something, the better teammate you become. Knowing that everyone can teach you something allows you to treat people differently. For example, if you walk into a room with the mindset that everyone is beneath you, you miss out on the experience of everyone there. Once, I was talking to our designer/user experience intern, who was my age. I expected him to have poor skills because he hadn't graduated yet and had no work experience. Boy, was I wrong! He had incredible skills! He not only knew about programming and could program quite adeptly, but he had also been a professional gamer, playing Starcraft 2 at 16! This gentleman was 22 and had done more than most 42-year-olds. I would never have guessed that an intern could accomplish so much in his life and could be incredibly skilled at his job. Experiences like these humble me when they happen, and I'm grateful that even though I go into every engagement with an open mind, I'm still routinely blown away. I've learnt some of the coolest things from the most unlikely sources. Always remember, you can learn something from everyone.

Rule Two: Respect Everyone in the Team

Whether it's the intern that's been there one day or the tech lead who's been there 40 years, you must respect everyone. This goes far beyond simply respecting them; it extends to listening to everyone's input, not talking over people when they speak, and truly listening to people when they say things. Nothing feels more disrespectful than not being heard. One of the most important things to keep in mind, with the abundance of notifications right at our fingertips, is putting it down when someone else is talking. I read a fascinating book years ago and have long forgotten the title, but it was about relationships. Ninety per cent of the book was about the idea of bids. What are bids? Bids are the little calls for attention, such as when your partner asks you what you think of the couch they are interested in online. Turning and facing your partner, looking up from your device, looking them in the eye and caring about what they say makes them turn towards you. By turning towards you, I mean they get a positive experience out of that interaction, strengthening your bond. When you turn towards them, they will turn towards you in the future. Now, I'm not saying you should treat your co-workers like your lover; however, I am saying you should treat them similarly, turning towards them whether they're telling you how their weekend was or telling you about a bug they have. Turning towards them, paying attention, and listening fosters positive interaction, strengthening your bonds and bringing the team closer. The closer the team, the better it performs.

Rule Three: Communication

In agile development, we have the concept of stand-ups, meaning we sit or stand and discuss what we have been working on, any blockers, questions, and concerns. This allows us to tell each other how we feel about the current work and how it's progressing. However, the communication often stops there, so instead of reserving it for five minutes at the start of the day, talk more often. Everything from personal issues that could be affecting your work to the back-end progressing so slowly that the front-end is left twiddling their thumbs. As the front-end developer, you could sit there, blame the back-end, complain, and get angry, but it won't help. Instead, talk to the back-end developers and ask why there is a delay. This communication can turn the hostility between the two parties into an understanding moment, which is why communication is key to any team endeavour. Having the ability to communicate effectively brings the team together rather than pulling it apart. The more you can bring the team together, the better time everyone has. The more you bring the team together, the easier it is to hit the targets.

Rule Four: Take Responsibility

If you write some terrible code, own it and ask how to improve it. Your teammates will be more than willing to lend a hand when you take responsibility for the atrocity you created. How many times have you written code, and it turned out terribly? Well, it often happens. Everyone has written some awful code at some point in their career, and it's come back to bite them. Just own it, don't try to hide it, don't pretend you didn't know what you were doing, and don't make excuses; just accept that it was terrible code and fix it. People don't care about the thousand excuses you give; they just care about how you plan to fix it. Accept that you screwed up and fix it. You earn more respect from the people around you by owning and fixing it than by pretending it's all good or making up a terrible excuse.

Being part of a team is one of the hardest parts of being a programmer. It requires a lot of practice; you must handle people's emotions and egos and deal with people who have

10 times your skills and 10 times fewer skills than you. You will screw it up and get into arguments, but you can mitigate it. You can be someone others want to work with and always respect the golden rule: don't be a dick.

BE A RUBBER DUCKY, FIND A RUBBER DUCKY

Every programmer will experience one thing in their career: being stuck on a problem. This could happen once a week, once a month or once a year, but at some point, it happens. Your initial reaction is to stay at work until it's solved. You would rather not stay, but you must because you've already been at work for 10 hours, trying to solve this all day. However, you should go home because there's a chance that the break will allow your mind to solve it or think of another way to solve it. Better yet, as you sleep, your brain will continue to work on the issue, and when you return to it in the morning, you could have the answer or another path to try. These techniques often work, but you've often just stared at the problem for too long. This is when your rubber duck comes into play.

In the book *The Pragmatic Programmer*, which I would highly recommend reading if you haven't done so already, the author talks about a developer who would explain what his code did, line by line, to a rubber duck he carried around with him. This process forced him to see his mistakes because by explaining it line by line out loud to the rubber duck, he picked up on the small mistakes that caused the issues in his code.

This can also be done by talking to another developer. There was a time when a developer in the team was having issues with his tests, the class he was trying to mock was returning a proxy object in Java, and it was racking his brain. He was vastly more senior than I, yet was stuck for a few days on the problem, so he bit the bullet and asked for help. In 30 minutes, I found the issue that allowed him to move on to the next stage.

I'd never dealt with it before, but my fresh perspective with no previous knowledge of the system allowed me to perform a few quick google searches that revealed information that solved the issue. A good senior developer won't be afraid to ask for help; discussing your issue with someone else or letting them look at your code can solve your problem. If your team supports it, you should spend some time talking through what you've done, how you understand the problem, how you'll solve it, how you did it, and what your thought process was. Talking through your code, even writing it out to send to someone, produces a greater understanding of it. Furthermore, it reduces the number of bugs and, most importantly, gives others in your team exposure to different parts of the system so that they have some knowledge in that domain.

Pair Programming

We have talked about showing your code to others when you have a problem, but what about pairing from the start?

Pair programming is the task of completing tickets with another developer for the entire duration of the ticket. At first, it appears to be performing the same work with twice as many people. This leads to the incorrect conclusion that the team's productivity is cut in half. This is not the case; quite far from it. In fact, the more you pair, the more productivity goes up when performed correctly. This is because, as a solo developer, you miss many mistakes, you get stuck, you might miss a return statement, spending 20 minutes trying to figure out why your code isn't working. With a second pair of eyes, these small-time sinks that add up are eliminated. Furthermore, as you get into the problem, you have a second person to bounce ideas off. You get stuck for shorter periods of time when you have two brains on the task. The best part is that the code review is almost redundant in the end, as another person has extensively reviewed your code. So, when it comes time to merge your code into main, it's less likely to break or have bugs because another developer has vetted it. This is tremendous for productivity, and as you can see, you can drastically reduce the number of developer hours for a task. You find that if it takes eight hours to complete a task as a solo developer, it might take three with a pair. Furthermore, the code the pair can output in terms of quality becomes drastically higher. There will be better tests and code, fewer bugs and more flexibility when it comes to modifying the code later down the line. Pair programming can be extraordinarily fun and fast-track your growth. Paired with a great person, you can become the best of friends. You learn an extraordinary amount, much more than you would by yourself.

Another advantage is onboarding people. It typically takes months to onboard a developer and for them to feel productive. Assigning them a ticket is a way to churn through many hours as they learn their way around. They do not understand where code is and how to utilise existing functionality to fit the bill. However, if you pair that person with someone with extensive knowledge of the solution, you can fast-track their learning. This drastically reduces the mistakes the new developer makes and reduces the number of hours spent on their early tickets.

When Not to Pair

You shouldn't pair program for everything because it becomes a waste of time when it's something truly simple. Contrary to solo development, you find it significantly more mentally taxing. You would be lucky to achieve more than five hours of productive work as a pair. This is because you are constantly engaged. This day after day, week after week can be incredibly taxing. The lack of downtime drastically reduces the number of hours that can be put in. This isn't all bad, as those five hours are much more productive than the eight you would perform solo anyway. You just have to remember the need for regular breaks throughout the day. It's all too easy to simply push through, losing all semblance of time, only to get to the end of the day feeling like that person crawling to a water source in the desert, desperate for a break, mentally fried. That's at least how I feel after I put in too many hours of pair programming. Solo development is also a way to relax, getting deep into a problem without that social interaction. You don't always want to be on, talking, discussing, and dealing with another person. You want to have time to yourself, focus, and sit in silence as you contemplate the problem. Pair programming is not the saviour of all issues. It is, however, a way to share knowledge, accelerate the development of complex work and flatten out the level of expertise in a codebase.

Pair programming is not popular because most managers believe you are cutting your productivity in half. Two developers doing the work of one appears to be a poor trade-off. If a team is not set up for pair programming, this might be the case, or at least initially. However, as the team becomes accustomed to it, they grow and progress rapidly. The sum of the parts is much greater than the whole, as they say, and this holds true when it comes to programming. The time saved on small miss-types or lack of understanding of a ticket is solved when you add a second brain. The two developers feed off each other both in mental appetite and focus. It is much harder to become distracted when you're working hand-in-hand with another person. As a junior developer, you will find this aspect of programming most enjoyable. Your growth and progression are exponentially compared to working by yourself. You have the opportunity to ask hundreds of questions. You have a second person to help you get unstuck when you become stuck. Pair programming is extraordinarily helpful even when you become a senior developer because you simply won't know everything there is to know. Even pairing with a junior developer forces you to explain concepts you might have taken for granted. This solidification of concepts helps you learn as well. Pair programming is a method I find extraordinarily powerful; it is something you should push for your team to undertake if you can.

WRITE CODE FOR OTHERS, NOT YOURSELF

Code and people are complex; therefore, code written by people is complex. When writing code, it's often thought that you write it to reach a goal. You have a feature you need to complete, and you don't stop until it's done. But how often do you take a second to ask yourself whether it makes sense to other people? It's hard and something I've struggled with for a long time. I can't tell you the number of times I've written some code, named a function 'abcdf' and had the function do 'fdcba.' I then change the functionality but forget to change the name, and when I push my code up, everyone reading it is dumbfounded, wondering why my code does one thing but is named another.

Naming code is one of the hardest tasks in programming. It requires description and detail but not too much since you don't want it to become too long. Furthermore, you must update it as you update the functionality of that function. However, it's not just the function names that can be a mess; it's the variable names too. I once worked on a project with thousands of lines of commented out code in production and variables named 'test' and 'test1.' The beauty was that neither of them was being used inside a test; they were used in production code doing god knows what. What is the point of naming these variables useless names? Laziness mostly, but also a lack of care for anyone reading your code in the future. Naming things is critical.

Documentation is something some people do while others don't. Some people believe it's a waste of time, while others believe their code is self-documenting. You might create a one-line function named 'getUser', and the function inside gets the user in one line. Okay, maybe you don't need documentation for that, but what if you call it 'getUser' when there are multiple or different types of users? Perhaps it doesn't fetch the user by 'id' but by shoe size. In that case, your function name is wrong, but it's also very unclear what the function does and requires when it's crammed between a sea of other code. When it comes time to implement a big feature such as payments, it becomes vital to document your code, plan what you're doing, and ensure everyone in the team understands what your feature is doing and how it works. This makes it easy when someone in the future has to debug it.

The last thing you should do before you submit your code for a pull request is to check that it makes sense to someone else. Have you named your functions correctly? Do they describe what the function does? Do your variables convey what they have captured? Does your return match what your documentation says? Does your function match what your documentation says? Does your code jump between naming conventions and/or ways of describing functionality?

You must take the time to read over the code as you would an essay. Proofread it so that others can understand it because if you don't, you will have a million questions coming your way.

As a company scales, they employ an intern program to test out graduates, put them under the wing of the senior developers and see how they go. Years ago, my previous employer hired a graduate, and he blew us away. In the first week, we tasked him with a relatively simple feature on the surface, but as you broke it down, it became more complex than anticipated. We broke down the feature, how it needed to work, what it needed to do, and possible ways to do it. At the end, we mentioned that he should plan it out, document it and get one of us to review it before he started to make sure he was on the right track. A day or two later, I received an email with the Confluence page for his plan, and it was incredible; it was detailed, easy to read and properly thought out. He got the green light to get started.

A few days later, when it was time to submit the PR, he linked the Confluence page again, only this time, there was also a detailed description of how the functionality worked as it had been built, why it had been done this way, what the trade-offs were at a particular decision point, and the benchmarks to justify it. Each function was tested and described, and, most importantly, it was easy to read. This intern was a pro from day one. When I looked over the code and was confused, I would go to his documentation and find the explanation. This level of documentation and attention to detail doesn't come around often, but it makes it easy to understand the code when it does. It also makes it easy when the author of the code comes back months or years later and doesn't have to sit there thinking, *who wrote this?*

When you have clear, well-documented code, you make your life and the lives of people around you easier. Yes, it takes much more time to document, test, and think about naming conventions and describe your code. However, the payoff is invaluable. It might take you 100% more time, but it will save you or others 1000% more time in the future.

TEAM DEPENDENCIES

One of the most challenging things you face in your career is the dependencies between teams. In a small company, this is fine; you can message Bob and ask for that latest API change. In larger companies, teams are typically broken into 10 or fewer developers who maintain a certain repo or part of the company codebase, maybe a few repos. If I need the latest API changes, I can still message Bob, but Bob might tell me they are waiting on the latest changes from another codebase. This is the nature of larger companies and large codebases. Many teams and people are involved, all of whom have deadlines and pressures. This can be enormously frustrating, especially when you are a customer-facing application with more dependencies the closer you are to a customer.

Take, for example, the addition of a date of birth field in the UI. Your team is responsible for the UI, so you create a ticket to add another field on signup that collects this information. In a well-designed system, adding a field to an existing signup form should take one day at most. Now the real complexity comes when you have to send that data somewhere. Your company might have a structure where its UI → Customer API → Internal API → Database. This leads to many dependencies if each of these teams has to individually edit their code to add this field or the capability to have this field. Herein lies the problem with large companies and many teams. To get this change through from start to finish, your manager tells the business it will take six months to add one single field to a signup form. Your manager explains the complexities of implementation, the teams, their deadlines, and the complexity of changing a database. Sure enough, nine months later (much longer than expected), the field is ready for users. In addition, it has gone way over budget and way over time. This is obviously a contrived example, but the point remains that the more teams you have spread out over more repositories, the more communication is required, and the more things can go wrong. Dependencies are a killer for any deadline. This doesn't even take into consideration that your team has been instructed to add one field, but now you have to force these other dependencies to complete work for you, which affects their deadlines, and projects and makes them angry. Again, this is a contrived example, yet I see similar things frequently happening in large companies.

What Is the Solution?

I have been thinking about this for some time, and I believe the only solution is to have teams that consist of people across the major dependencies. Microservices have numerous benefits, and often, microservices are done in many repositories; having a team that can access each of the services to add their code removes the dependencies. Instead, the team that needs the feature must get it done; they can't simply throw their hands up and say they are blocked because the team didn't do the work. Instead, you create teams of people from all parts of the organisation. You could have a few front-end developers who have worked on many front-end repos, or at least enough to get the user DOB field in. Then you have a developer from the 'Customer API' team, another from the 'Internal API' team, and lastly, a database guru. You have one developer from each section or dependency you would have previously relied on. What would have been many teams is now one, and they all have control over each part of the application. Thus, reducing the number of dependencies reduces the team's pain. In this manner, you could reduce the time from the eventual nine months to a month (or a significantly smaller number). You have fewer communication issues, you have someone who knows about each part of the stack, and each person has the same deadline!

The business delivers features faster, people still get to specialise in the area they want to or try new things, and there isn't a communication issue. Put simply, having more teams increases the amount of communication that must happen, but how you split these teams up determines if it will be a pain or a pleasure.

Companies that have many teams in silos always produce toxic results. Teams hate each other; in fact, it can get downright nasty when one team relies on another to get the work done. This results in both teams not working well and generally creating an awful workplace. No team wants to produce bad software full of bugs and delivered late. Teams aim to do their best work; they just need the organisation to support this. This, too often, becomes the norm, allowing poor culture to grow. I have seen people have breakdowns because of the stress of it all. This might have been one of many reasons, some of which are personal, but the point remains: if you can remove these kinds of stresses from the workplace and increase the time to market, why wouldn't you?

MARKETING YOURSELF

Set up a Website

One of the best things I did early in my career was to set up a website. However, I didn't make a custom one from the ground up; I just created one on Squarespace. You can use any hosting platform, but I recommend one that is yours, one that allows for your own domain, to create your own brand. Complete control over your website is a must; you should have pages for projects you have worked on, a blog and an about section. One thing that I like to have is a 'friends' section. It sounds funny, but it's a great way to build a network. When people check out your site, they can see others you know personally and vice versa. In my case, I met somebody early in my career who became my mentor, and we kept in touch and became great friends. He also had a website and was much further in his career, 10 years further, in fact. This meant that having him in my 'friends' section would mean that those looking at his website could potentially come to mine, and those who looked at mine could see that I worked with some smarter people than myself. It's a way to show you socialise and interact with others in the industry.

Whatever platform you use for your site, make sure it's easy to set up, easy to maintain and something that can track analytics. It must be clean, work on all device sizes, and not spam people to sign up for a newsletter. This is why I didn't write my site because I would then have to maintain it. Fixing small CSS or layout issues is just not something I have time for when I just want it to work! I can easily copy and paste my article to the site text editor, add some images, make sure it's spaced properly, hit submit, and it's out there. If I'd like to, I can schedule the article to come out at a different time in the future. It's these simple features that make my life easier. Some might say you can't be a developer without creating your own site. Then they spend hours building and maintaining that, which is fine if that's what you want to do or if you want to have it as a portfolio piece. At the end of the day, an employer will be more impressed with your writing on software topics than if you can put together a website by yourself.

For one of the jobs I applied for, I pushed to get a higher title and more money. The thing I felt impressed them was the fact that I had a website. I don't have an exceptional site, and it didn't take much to create, but it pays considerable dividends. Furthermore, if you want to do some freelance work, it's a great place to send people. It's just another piece of evidence to show you know what you are doing.

Post Code to Places Like GitHub

This is not something I do that often, mainly because the code I'm writing is just for learning, but I would still encourage you to publish as much code as possible or contribute to a project. It becomes another layer of credibility for getting a job or promotion. The more you build your reputation in the industry, the easier it becomes to meet interesting people and get raises and new jobs.

Another benefit of posting your code for others to see is that people might suggest improvements, or you can reference it in your article to show people how to do something. There might be something in a new framework that you found tricky. Write an article and show some code of it; it could help a few people or maybe you in the future when you inevitably forget how to do it.

Write

Write, write, write. As much as you can, write and post it publicly. People will eventually read it, and when they do, they might learn a thing or two, which is what you should strive for. Teaching others is rewarding and enables you to solidify your understanding of a subject. It takes a lot of thought and work to compile your understanding of a subject into a coherent stream of words. You must understand what you are talking about and articulate it in a way that easily brings people into the subject and expands their knowledge.

It's also fun. You can take a topic and dive as deeply as you want, provide diagrams, code examples, or simply write about topics and your thoughts on them. Let your mind spill out onto the page. I don't enjoy the time spent preparing code examples and images, but I love writing. I will typically write about soft skills such as teamwork, code review, ethics, etc. It allows me to think deeply about a topic or problem and provide some information about it that hopefully causes someone else to think or learn something new.

You don't need to write every day or every week; you can do it as you see fit. I go months without writing anything, and then I might write three or four in one go. The most important aspect is simply having it there for others to see. Even if no one reads or sees it, you still go through the process of formulating your ideas into a written form, which is crucial for your job as a software engineer.

Have a Living Resume on Your Website

My favourite part of my website is the living resume because it shows off the projects I have worked on, who it was for and a summary of them. It's short, to the point, and highlights what I've done at a glance. You don't need much, just the projects you have worked on, both paid and unpaid, but keep only the most impactful projects, don't include everything. At first, this might be hard or impossible if you haven't had a job in the industry yet, but over time, you can add them one at a time and watch your resume grow. The hardest part about the living resume is elaborating on exactly what you worked on without making it too wordy. It's not easy; it takes time and many rewrites.

Share your website around and send it to all your job applications.

It's impressive when someone attaches a website to their application. When you look at two candidates, and they are as close as possible, the one with the website stands out. It just shows a little extra. Share your site with your friends, family and places like LinkedIn. It might be embarrassing to share it, but it's worth it because everyone will provide feedback to help it look and feel better.

Do a Talk or Multiple

It's nerve-wracking to do a talk, but it feels incredible once it's done. Pick a topic that interests you, dive deep into it, and present it at a local meetup. The skill of talking to a group of people helps you tremendously in your career. You progress faster and farther than others simply by being able to talk to people higher up than you and a large group. If you don't have many meetups in your area, do some talks at work. Many companies have brown bag sessions, which are lightning talks of 5-10 minutes, where you can discuss any topic you like. Others have full-blown 45-minute sessions where you can dive deep into a topic. It's important to do these as often as possible, even if it's simply to flesh out an idea and get some experience talking in front of a group. The further you go in your career, the more you need to present to others and the more confidence you need to get your ideas across. Become that developer in your company that is the expert in an area, take that interest in some area and make it what you are known for. I love Elixir, so everywhere I went, I would do a talk on it until I became the 'Elixir guy.' Performing a talk in front of a group of people is helpful in your career.

I put so much emphasis on marketing because you can be the best programmer around, but if no one knows you exist, you won't get much work! It doesn't have to be one of your strengths, either; it can be a simple website with simple articles on a pre-made platform. Whatever you do, make it simple and do it well.

SECTION 4

The Rest

TAKE BREAKS

This is something I am not good at; I go months without a single day of leave. However, I shouldn't, but I do! I get stuck expecting deadlines all the time. It becomes an endless cycle of not taking leave, pushing it off, and never taking it. Don't do this; instead, take regular leave, even if it's a long weekend or a few extra days around a public holiday. This way, you can take advantage of some time off without using up all your leave. Four weeks of leave isn't enough each year, especially when you factor in that most companies have a shutdown around Christmas, meaning you have to spend a lot of your leave there. If you can afford to take unpaid leave, try. You become extremely burnt out and tired if you go months or an entire year without leave. You hate your job, struggle to do anything, and become grumpy. Taking time off to do something else will get you back in the saddle again, loving life and your job. I start hating a job after a year, typically because I haven't taken any time off that year.

Take breaks to reduce the accumulation of stress in your life. You won't realise how much you have missed time off until you have a four-day weekend and can't wait to have another one when Monday rolls around. It's important to plan a few trips in the year; they don't have to be big but something to look forward to, something you book months in advance. The reason you need to book these in advance is that if you don't, you won't take them. If you are like me, things at work will come up, and you will never book your leave. However, it's important to remember that taking breaks doesn't have to mean taking a break from work; it can mean a break from learning or side projects as well.

Working Less

There aren't many fields that allow their workers the flexibility to work fewer days, but if you plan your finances, you can work less than the traditional five days per week. You might find working four days suits you much better than five, so take that extra day off, take the pay cut and plan your finances. I know someone who is incredibly smart yet only works four days per week because he wants to avoid getting burnt out. You can also follow this path if you desire. Of course, you will get paid one day less, but you gain more flexibility in the process. You can use that day to recover, build a side project, spend it with family, etc. The freedom that a development career can give you is extraordinary.

Working from Anywhere

You could take a break four weeks per year and do the traditional holiday thing. On the other hand, you could work from anywhere in the world if you choose. There are plenty of people who travel the world and work from anywhere. Take a laptop and an internet connection, and you can work from anywhere. You can travel all over the world and never miss a day of work, or you could spend fewer days working and travelling more. I have also heard of people working six months per year to then travel for the other six months. It's all up to you and what you want to do. You can work five days per week, 52 weeks per year for 40 years, or you can work the minimum to survive and live an alternative life. You have the power to pick any path you wish; you just have to have the courage to live it. In today's age, you don't even need a powerful computer. As long as you have a good connection (4G), you can connect to a service that provisions your computer for you, or you can connect to a much more powerful machine you own somewhere else. I once read an article on someone who would program off their early-generation iPad. They connected it to a server they owned, using that as the main hub of their work, and it would perform all the heavy lifting; all the iPad had to do was be a keyboard and a screen. They travelled around the world, lived from a van and had the time of their life. With remote work being a bigger option the farther we go into the future, the more chances you have to work from anywhere in the world. Gone are the days when you would need to take four months off to travel or wait until you retire. You can do it now, and you can do it while you work.

GET A MENTOR

I stumbled upon my mentor early in my career, around the 13-month mark. I was working for Racing.com at the time, doing work to migrate schemas between two DynamoDB tables to help with some testing. Furthermore, I was assigned to the lead solutions architect at the time, Andrei, who showed me what I needed to do. I was only on this project for a month at most. However, during that time of daily lunches, chats and working with him 9-5, we became friends, and as he had 10 more years of experience, he became my mentor, whether he wanted to or not. We still talk regularly, and I still go to him with questions and ask for advice on various topics. Andrei has helped me fast-track my career simply by being a friend and giving advice when I would inevitably complain about a situation at work.

There is no way I could have gotten as far as I have as quickly as I did if I didn't have a mentor. A mentor doesn't always have to be someone with more years of experience than you; they could just have more experience in certain domains or technologies. I have a friend/mentor who specialises in a certain domain, so when I ask him questions, he often has the answer. We have roughly the same number of years of experience, but he has an entirely different skill set and comes from a different background. Something I have come to love about software engineering is the sheer diversity of people's backgrounds. Gone are the days when you needed a university degree to get into the field; you can get into the field with as little as some hard work and a 12-week boot camp. I have seen people with law degree backgrounds who have done exceptionally well. The background in negotiating and communication skills you get from a law degree puts you far and above the rest when debating a solution. A computer science degree does a fairly average job of preparing you for the real world; it gives you some skills to start your journey but omits a lot. One of these things is communication, something you require in your career. Getting a mentor exposes you to these viewpoints and people with vastly different backgrounds and experiences. These people become your friends and help you grow into the professional you are destined to become.

Why You Should Get a Mentor

Mentors are like cheat codes in a video game that allow you to input a code, and suddenly, you have everything unlocked. The right ones are that good! I have learnt an incredible amount from my surrounding mentors; they have taught me how to handle situations or what to look for in a new job. In fact, one of the best pieces of advice I got early was from Andrei, who said, 'Don't be loyal to a company; if you want a promotion, and you don't get one, leave and get one. It's much easier to leave a company to get a promotion than it is to stay and hope you get one.' This stuck with me, and once I hit that 1-2-year mark at a company and didn't see a promotion or any opportunities for personal growth, I left. I don't feel guilty or lose sleep over it; I just look at what is out there, and if something pops up that fills my criteria, I take it. It's one of the secrets that helped me progress through my career quickly.

You might be able to get some advice from books, but at the end of the day, the people around you give you much better advice than any book. They see you daily, know your weaknesses and strengths, and hopefully, are honest enough to tell you when you are being terrible or when you need to stop complaining and do something about your situation. A mentor has been through most of your problems, and they have had mentors themselves. They have solved problems you are trying to solve now, and they can give you quick and concise answers to solve them and what to do in a situation.

How to Get a Mentor

So, how do you get a mentor? Well, you need to be social and talk to people. Chat with your tech lead, the solutions architect on the project or the CTO of a start-up bumped into at a local meetup. Perhaps you got lucky enough that your company hired an expert consultant in a domain. Chat with these people, and if you click with any of them, keep talking. I don't have any mentors that I wouldn't have as friends. Sure, you could pay someone to be a mentor, which might be the separation you need, someone who can give you brutal feedback. However, not everyone wants to pay someone or can. Most people are willing to help if you are eager and seek their advice. Most of all, listen when advice is given. I love when someone approaches me with problems and asks for advice because I have lots of technical experience, and I would say others do too. Helping someone and giving them advice that you wished you had or found amazingly helpful in your early days is rewarding. If someone is willing to help you, you won't have to force it, and if you click on a personal level, you become friends, and the whole process will be natural. I don't consider my mentors to be mentors; I just see them as friends who just so happen to have valuable experience in areas I don't have or haven't gotten to yet.

Okay, what if someone bites your head off for asking so many questions and being annoying? That's a real fear, but it's a problem; some people simply don't want to help. These are people to avoid, but you must get over the fear of this happening and ask for help regardless.

Team Mentor

You might have a mentor in one company or team, and you no longer speak to that person when you leave that team. This is also fine. Having one person you can come to with any problem at work is something truly special because it means there is one less person who could potentially be lost in the team.

You might find that the company you are working for has a poor culture or the time pressures mean no one is available to help you. Leave as soon as possible if this is the case; they don't deserve you. No matter what, never compromise on your value, learning, and future for a company. Any company that doesn't help people who need help is not one to work for. Occasionally, this might be a single team where the senior members are not leadership material or simply don't like helping others. In this case, leave the team or company if you can. This is your career; if someone in your team can't be a mentor, it will be a horrible team to work for.

As the lead in the team, I take personal responsibility for anyone who isn't happy. I go out of my way to fix that. It isn't worth it if the project gets delivered, but the entire team suffers. The aftermath of that release will be horrible; people will leave, get sick and become less productive and miserable. Your tech lead should be the mentor in your team; you don't need to tell them your personal life stories or be friends with them, but you should feel comfortable to go to them with any work problems, and they should be able to listen with empathy and help accordingly.

The tech lead is the ship's captain, and you, as a developer, are a crew member. Ultimately, if the team succeeds, it is because of the team, and if the team fails, it's because the captain failed them.

BE A MENTOR

We have discussed finding a mentor, their value, and avoiding people who don't want to be one. However, we haven't talked about being a mentor. You don't need 20 years of experience to be a mentor; you simply must be willing to teach others. Everyone has something to teach, no matter how new to the software they are. You know things that others don't, but it's not for lack of learning; it's simply the lack of experience someone has. No two people ever have the same career experiences, which is what makes everyone so different and why you can learn something from each person, no matter their level. You know a lot about a topic, and you won't even realise it until you start to teach someone else.

When I first became a tech lead, I was terrified that I couldn't teach anyone anything and that everyone on the team would know more about programming. However, I was shocked at how much I knew after the first week. When developers asked me to look at something, I could diagnose the problem or provide multiple paths for them to try. I knew much more than I cared to admit, and thankfully, it all shone through when I needed it most. You surprise yourself with how much you know when you start teaching others or diagnosing other people's problems. It's even sometimes just about seeing things from a fresh set of eyes that is enough to solve the problem.

Whom to Look Out For

Who should you mentor, and what should you look out for in a mentee? Find an eager learner who asks a million questions and is downright obsessed with programming. They need to listen to what you say and be hardworking, not someone who just wants answers, so they don't have to do any work. You must also get along with them, as there is no point in being a mentor to someone if you dislike them. More than likely, they won't ask you to be a mentor. Offering advice or chatting about tech in a conversation that flows naturally is the best kind of mentoring because it feels more like a friend offering advice, which they can take or leave at their will.

Having a mentee shouldn't be a forced process, and you shouldn't be on the lookout for one; it should just happen naturally because that's when you know it will work.

You don't have to maintain a constant connection with them. You might not talk for months at a time, and then suddenly, they come to mind, and you reach out. The more you allow it to flow naturally, the easier it becomes.

What Are Your Responsibilities?

As a mentor, you are responsible for giving your mentees good, sound advice based on your experiences and learning, offering help when possible and being on the lookout for opportunities that might fit their needs. You might have a friend looking for a developer, and your mentee would fit the role perfectly. So, you get in touch, and likely, your mentee gets the job based on your reputation with your friend. Your mentee might not even have to interview, so it's easy for them, and everyone wins. You likely received these opportunities from your mentor back in the day, so you must pass the baton to your mentees. Think back to when you got a lucky break from knowing someone who knew someone or a senior developer who sat down and helped you through a ticket. You can do that for someone else; you can pass the baton of opportunity onto the next person who needs it.

Everyone needs a mentor for different purposes. Some people need a life coach to get them in order, and others need someone for an aspect of their business. Everyone requires some form of a mentor, but only the small few deserve one. Most people won't look for one, and most people don't even want advice, so refrain from giving advice unless specifically asked. The most annoying thing you can do is offer unsolicited advice; the only time you can do so is by asking them if you could give them advice. Often, people don't ask for advice or help because they are afraid to ask, so it's important to ask first.

However, you must remember that as a mentor, you are not there to give your mentees all the answers; you are there to guide them through the path you have already taken. You are a shepherd to a flock of sheep, guiding them through the hazards. When they come to you for advice or a problem, it is up to you to draw from your experiences to offer advice. Often, you push them to toughen up, deal with it and put some things in place to better handle the situation in the future. Frequently, I find that juniors simply haven't had these experiences and thus feel everything is crashing around them. When deadlines arise, management typically puts a tonne of pressure on people to hit them. When you are new to the field, you see these deadlines as the be-all and end-all, and the pressure gets to you. As you grow and deliver many projects, the deadlines become meaningless because they can always be pushed back. If you are the tech lead in a team, it is your responsibility to shield the team from these pressures while highlighting the importance of the deadline. Furthermore, you must offer advice and reassure the team that they are doing their best.

I Don't Feel Ready

I didn't feel ready going into my tech lead position. Furthermore, I thought I knew nothing. However, when you feel you aren't capable of being a leader on a team, the more capable you are. If you are overly confident that you can do it, you're probably overestimating your abilities. You should feel that you can't do it because it means you understand that you have much to learn, and when you are open to learning, you discover that you already know a lot. So, when you think you're not ready to be a mentor, you probably already are ready.

You Don't Have to Be a Mentor to Everyone

Becoming a mentor to someone doesn't mean you have to become a mentor to everyone. You're only human and, thus, have limited time and capacity to help others. You should limit who you mentor to those who would benefit the most and those who deserve it. Being a mentor can be extremely hard, so you should limit it to only a few people at a time. I would say you can only mentor five people at most as this is a small enough number to remain personal with everyone and not too big that you spend your entire life mentoring people.

If you enjoy being a mentor, you should consider having people pay for your mentor service. It is a great way to filter out the people who truly want it, but you should still be mentoring at least one person for free; it's all part of giving back to the field.

Ultimately, you probably only mentor a few people since most people don't look for mentors. Looking for a mentor requires vulnerability since you're putting yourself out there. Most people can't be vulnerable with others; they simply don't care to progress or believe they know best. These are all valid reasons why most people don't get a mentor. Thus, it is up to you to choose those who want to improve.

TAKE CARE OF YOUR HEALTH

The Importance of Health above Everything Else

I have been extremely unlucky in that my family have been affected by health conditions that either took their lives early or seriously reduced the quality of their lives, sometimes to miserable levels. Poor health is awful; you don't realise how good you have it until you have an ailment.

Not every health condition can be solved by taking care of yourself; you might just get unlucky. However, you can significantly reduce your health issues by paying attention to your health as early as possible. It doesn't take much; it's the 80/20 rule, where 80% of the outcome will come from 20% of the work. This chapter will not be an in-depth guide on what to eat, how to eat and how to exercise, but it will be a general overview of health.

I was fortunate enough to play sports, eat well and have family members that taught me the fundamentals of weight-bearing exercises from an early age. These influential people showed me the importance of doing things right and not rushing. Anyone can exercise, but doing it properly to perform these movements late in your life is something most people do not think about. I was incredibly blessed to grow up with an uncle who was an amateur bodybuilder. He taught me the fundamentals of performing exercises with the proper technique, slowing down, and thinking about your body as a vehicle to maintain your entire life. You can lift weights six days per week and create an astonishing physique, be in top physical shape and still have awful technique, which will eventually come back to bite you. You want to avoid getting to your 50s and needing back surgery because you never learnt the proper techniques to lift safely and effectively. I say this to warn you that you need to take care of your health, but that doesn't just mean doing anything in a gym or running 10 km per day. It means finding the physical activities you enjoy doing, which can be anything, and then performing them in a way you can see yourself doing late in life. When exercising in a non-contact sport, you should not get injured regularly, feel physical pain (unless it's sore muscles), or damage your back. You should perform controlled, slow movements with weights appropriate to your level. You don't need a lot of weight; you just need to load your muscles and bones. Why bones? Well, it has been shown that elderly people who perform even light weight-bearing exercises significantly improve their physical body. Weight-bearing exercise fights off osteoporosis and allows your body to do what it needs. You don't want to be 80 years old, fat and unable to move because you never exercised; all your bones and muscles will ache, and you will struggle to do anything. By performing a small amount of weight-bearing exercise, your body will keep building those muscles rather than wasting away as you age.

You also don't want to be someone who yo-yos their weight, constantly losing and putting on weight, much like a boxer trying to cut weight for a fight. This weight fluctuation is enormously destructive for your entire system and will, over time, make it harder to get down to that healthy weight. You want to strive again for that 80/20 rule; eat well and exercise well 80% of the time, and the other 20% will be okay to have a muffin or McDonald's. Life isn't about restricting everything so that you can live the longest and most boring life; it's about enjoying the years you have in the best possible health you can have. I have seen first-hand what poor health does to you; it might take 50 years, it might take 70, but eventually, it catches up, and it hits you like a tonne of bricks.

My grandfather was a navy sailor, who became a detective in the police force, and later became a bodyguard. These are all physical jobs, and he was fit for his entire life. He ate what a standard 20th-century diet would have been, meaning less processed foods, but he also smoked all his life. When he retired from being a bodyguard, he did so years before his wife. With the trauma of working in the police force and the boredom of retirement, he took up drinking and smoked more. He smoked and drank for over 20 years to the point of alcoholic dementia, and he died by 72 from complications of alcoholic dementia and emphysema. He drank and smoked his entire life, but it wasn't a problem until it caught up to him when he got older, and it was too late. Do not be like this; take care of your health now to stay on top of it for your entire life. The four pillars of health are diet, exercise, stress, and happiness, and you need to keep these pillars in equal balance to have a long and successful career and life. Letting any of these slip will result in poor health later down the line, and it's not a matter of if but when. I often see obese people in their 50s and 60s, and their life is a struggle each day. From getting off the couch to walking around the supermarket, their day-to-day life is physically painful, and the regret of not taking care of their body weighs heavily on them. It doesn't take much work; it just takes consistency each day to build the proper habits for a happy, healthy life.

Exercise

The earlier you start taking care of yourself, the easier it will be. It's just one big habit; the longer you have a bad habit, the harder it is to break. If you smoke for 40 years, the effort to break that habit will be magnitudes more than if you only smoked for 40 days. You don't need to take massive leaps, drastically changing your lifestyle; you just need to make small adjustments at regular intervals. It's about a slow lifestyle change rather than a drastic, dramatic one like most diets. Of course, this all depends on how good or bad your lifestyle is. If you never exercise, start walking 20 minutes daily on your lunch break. Getting out into the fresh air, even when it's cold, will drastically help your mental energy and clarity, and you get some low-grade, easy exercise that eases you into better health.

If you hate exercise, that's completely fine, so build a habit of walking, not for the exercise but for the mental improvements and happiness you will receive from being outside and away from computer screens. If you hate exercising, you need to learn to love something else about it. You need to learn to love the improvement to your happiness that you get from spending 30 minutes walking in the sun and fresh air. As a bonus, you also get in 30 minutes of walking.

The more exercise you perform, the better, but at a minimum, you should be doing 30 minutes of walking each day. This could be on your morning commute to the office or during your lunch break; it doesn't matter. Just build the habit each day and do what you can. I don't want you to go out and do 10 km runs if you hate running, but if you don't mind it, start small and then build up over time. I hate cardio, but I love basketball, so by playing some basketball, I get my cardio without thinking about it.

Weight-bearing exercise is more important than cardio. Studies show that in old age, having a regular weight-bearing exercise regime will give you a greater quality of life with more mobility and reduced osteoporosis.

Gym

If you don't mind gyms, go to them. They have everything you need; you can do an entire full-body or upper-body/lower-body split in 20 minutes and get all the benefits you need. If you hate gyms, and frankly, I hate them too, you can get some adjustable dumbbells that allow for multiple different weight settings, a bench if you have the space for it and a kettlebell; that's all you require. You can build a great physic and be the healthiest version of yourself from 20 minutes of working out at home, done correctly with the right form. Keeping rest times between sets low also helps keep your heart rate up, as cardio would. Personally, if I haven't worked out for a while because I was sick, for example, I need to take it extra slow, so I might only do five minutes to start with, then slowly add one exercise at a time. I prefer to do five minutes daily, over 20 minutes three days a week. Furthermore, I also find that the physical toll, at least in the early stages, is quite high; thus, if I go too hard too quickly, I burn out. Then I'm right back to square one. So, with that in mind, start slow, and make it a daily rather than a weekly habit. It becomes much more ingrained when it's daily.

I Don't Know What I'm Doing

At some point, we have all had to learn something through trial and error and the guidance of others. I was lucky to have an uncle who taught me the basics and most important lessons of exercise. These are:

- Slow and steady repetitions
- Strict and proper form
- No compromise on form over heavier weights

Taking these three guidelines as gospel allows you to research the exercises you want to do, picking one per body part (if doing full body each day) or two if you are doing lower-body and upper-body split. After you research how you perform these exercises, how much weight you should use and how many repetitions you should do, you will have a list of exercises you can try, adding those you like and removing those you don't.

Start by taking it slow to avoid injury, and use weights you can handle. Furthermore, you will get sore, but this is a sign that your body is adjusting to the program, so take your time. If you need more rest for the first few months, that is okay; take it. I highly recommend starting with full body, taking one exercise for each body part if you use dumbbells and kettlebells. If you are lucky enough to get a squat rack, which you can use for almost anything, I would recommend the four main compound exercises. These are squats, deadlifts, bench presses and military presses (overhead presses). These are more complicated lifts and sometime require a spotter or a teacher in person to correct you. If you can do these, you won't need to do anything but these to be extraordinarily strong and healthy. You can target just about every muscle in your body, and when done with the appropriate weight, the chance of injury is much lower than in isolated exercises. You could finish this workout in 20 minutes and feel fantastic for the entire day. Be warned, though, no matter how light you go for the first couple of times, you will be sore. You might have to start with one day a week, then add one more day each week until you get to 5-7. I'm not a fan of rest days unless you need them. I believe you should be doing little enough work that you can do it every single day. This way, it's an easier habit to maintain if there are no days off. Furthermore, I find that 15-20 minutes per day without pushing myself too hard is enough, although as you get used to the program, you might need to up the intensity by taking shorter breaks and performing more work. I don't find that spending more than 30 minutes on weight training is beneficial because it results in spending more time with breaks between sets. Instead, have shorter breaks; thus, you can up your intensity without extending the time and taking up more of your day. I will never understand the point of spending two hours in the gym and sitting between sets for minutes doing nothing but scrolling through your phone. You could do the same work in 30 minutes with 15-second breaks in between.

What if I hate walking or the gym? Try a sport, even for the social aspect, once or twice weekly. One hour of sport per week is good for you, both socially and fitness-wise. It doesn't need to be that physical either; it just needs to get you to move. However, it needs to be somewhat enjoyable so that you keep doing it. If you hate the gym or hate the thought of dedicating a set amount of time each day to doing something just for the health benefit, then picking a sport is the best thing you can do. You might have been a good swimmer as a kid, so pick up water polo. It will be hard, but it will be social, and you will have lots of fun. Maybe you liked football (soccer) when you were young, so get back into it, get some friends or join a local club. Football is great because you typically have a structured practice session, which is another hour or so that you get exercise in while being social.

Even if you have never done a sport, find something you might be interested in. Give it a try; you might like it or hate it, but at least you are trying something new, getting active and becoming fitter without actively setting out to be.

Diet

So many diets exist, but what works for one person might not work for another. What you can ultimately count on is that sugar is generally useless for your body; the more you have, the more you crave, and the longer you have it, the harder it becomes to reduce. Take, for example, if you had one kid who had less than 10g of sugar each day for the first 20 years of their life and another who had 100g a day. More than likely, the kid with 100g of sugar a day is overweight, and if they aren't now, they will be. Now, say that both kids are overweight, and they try to improve their diet. Who will have the most cravings for sugary food? The kid who consumed more their entire life. Why? Because their microbiome, which are the bacteria in their gut, are built to digest sugar. Therefore, the more unhealthy food you eat, the more your body adapts; thus, the more you can have. The less of these foods you have, the less your body can handle. Removing bad foods from your diet is difficult because your microbiome drives you crazy as it dies off. The bacteria don't want to die; they release chemicals to tell you to eat bad food to keep them alive, which is why you get cravings! We all get cravings, and we all have had bad food in our lives, so we all have some bacteria that are accustomed to this food.

However, it's not about changing your diet to lose weight; that should never be the goal. The goal should always be to improve your overall health for the long term. Everything you do should be for the long term, so the question is whether your 80-year-old self will thank you.

I don't write this advice as an expert on health and diets; I write it as someone who is interested in my health and has seen the effects of a poor diet on others. I don't expect you to follow any specific diet; it is up to you to determine how you want to eat. I write this to convey that you need to plan how you eat because you won't be able to eat poorly for your whole life without serious consequences. If there is anything to take away, it would be to reduce your sugar and processed food intake as much as possible and increase the amount of home cooking with unprocessed foods. You can follow any diet with this model of thinking. However, you need to think about how the food you consume will affect you because every calorie you consume will affect you somehow.

Stress

Stress is a factor in health ailments, but you don't see it until you or someone you know gets seriously ill because of it. We don't see the slow drain stress brings as we age and slow down. Our body gets accustomed to the constant state of stress, the flight or fight state of our nervous system because that is what constant stress does. I used to want to create a massive business and earn a lot of money. Then I saw how the stress of running one impacts those who do it. I didn't want any part of that. No figure in the world would make up for 10, 20, 30 or 40 years of daily stress and the constant fight or flight response your body must handle. Don't get me wrong; you can put up with it in the short-term and maybe even in the medium-term but one day, like your diet, the stress of your life will catch up to you. When it does, you will suffer from a serious health condition. You must get your stress under control in any way you can, and the first step is to reduce your exposure.

Reducing your exposure to stress is one of the most effective, but it can be the most difficult as it means removing the external factors of life causing stress. You might work in a consultancy that requires you to work crazy hours to meet deadlines, and as a result, you are stressed to the max. First, don't work more than eight hours each day because your potential career advancement isn't worth it. The company should look after its employees so they don't have to work this hard. Secondly, you can maintain that level of work for a short period of time, but not forever. Again, this type of lifestyle raises your stress levels. To avoid this, switch jobs or make a hard rule not to work longer than eight hours. The level of calm you get from a job that you can switch off after eight hours is truly spectacular. You can work harder and more efficiently when you trim down the hours and reduce stress. Some people on social media tell you to hustle and work ridiculous hours to make your dreams a reality, but at the end of the day, you may or may not get there. If you do, is it worth it if you're stressed, have chronic health conditions, or have a terrible relationship with your kids and partner? No, you don't hear people on their deathbeds wishing they had worked a few more hours in the office. They wish they had spent more time with their kids, watching them grow and teaching them how to be great people. Say you don't have any family or friends to worry about, and you want to hit the top levels of the development leader board, well, you can do that, but you will probably be miserable and unhealthy as a result. Some might say they'll just retire after that and will be fine. Unfortunately, their health will have already suffered by then, or they might be well past starting a family and settling down. They are left with the success of their former self and a lot of money, but they won't be happy.

Happiness

Happiness makes you feel good, but it's something you usually put off until the weekend. You work 40-plus hours each week and leave the happiness to the weekend, then you do this for 40 years, and then you can always be happy. Living only for the weekend is quite a miserable way to live your life. Perhaps you love your job, and that's great, but you still need to be happy outside the bounds of your work; you are not only your career. You need to be truly happy that you have the partner of your dreams and that you are on the path in life that you truly want to be on, not one prescribed by society.

You must spend time thinking about what you want in life. Is software engineering something I do for a job or something I would do in my spare time and for free? It also doesn't have to be the only thing you do. I would encourage you to have hobbies and a life outside of software engineering because it allows you to cultivate your happiness, rounding out your skills and interests. The key to happiness is thinking about what you truly want in life and aiming towards it while maintaining a happy balance. It's fine if you would rather not be in the top 1% of developers. You can happily enjoy the perks of being a software engineer, doing something you enjoy, but ultimately, it's still a job for you. If you love programming and would do it every minute of the day, then why not spend every minute on it as long as it brings you joy and doesn't affect the other pillars? You don't need to work a job 9-5 to do it; you could save, invest and retire early to pursue whatever programming you want to do. Or maybe you just love the company and your work, which is great and builds up your happiness pillar too. Life is about being happy. You don't need all the money in the world to be happy; you just need enough to live and develop the freedom to do what you want.

Get a Standing Desk

I never thought a standing desk would be beneficial until I used one at work. The ease at which you can change positions, stand up and stretch your legs is remarkable! It gives you an energy boost when you get up and stand during the day. I don't spend more than a couple of hours standing up at my desk each day, but it's enough to help my back and gets me moving a little. You don't need to break the bank, either. You can get relatively inexpensive standing desks that hardly cost more than a cheap desk, and they do the job perfectly. I didn't think I would use a standing desk that much, but it drastically changed how I worked when I got one. It's not something that needs to be done all day, every day, but just a bit each day. You will thank me when you can stand for a bit after hours of sitting down yet continue to work.

CONCLUSION

What I Learned

I learnt 100 times more than I would have ever imagined, from how to break production to functional programming languages. I learnt how to work in a team, how to work in a remote team and how to deliver a project to production. I performed production support, worked weekends, and worked a lot. You will learn an insane amount in your first year, more than any other year by far. You soak up knowledge like a sponge, asking a million questions and getting stuck on just about everything. I learnt that everyone is interesting; you just have to listen. I discovered the joy of delivering a project from idea to being in the customer's hands. Furthermore, I also discovered the stress, frustration, and hardship that is software development. There is always more to learn, and much like the medical field, there are many specialties that you can go down. You don't have to master them all, but you have to master being a professional software developer and create code that is easy to read for others and can be easily tested, deployed and refactored. I learnt that complexity is the root of all evil when it comes to code.

What Can You Take from This?

I wrote this book because there are few books about soft skills, the experiences you might have and the pitfalls you will absolutely live in software engineering. For this reason, I wrote this book so that, as a newcomer to the field, you would have proper expectations of it all. I had little knowledge of what it would be like and knew no one in the industry apart from one junior who had a job for six months. This book is something I wished I had when I started so that I could get a better jump on my career. In summary, if you're just starting out in your career, you should now have the knowledge to get a head start. You should at least understand the importance of being a professional, writing clean code, learning from others, and mastering your craft and the career ladder. Ultimately, this career path is great; you might start here, but you don't have to stop in this field. If it's not serving you, move on, take what you have gained from this incredible field and apply it to whatever you wish to do next.

What I Would Do Differently

I wouldn't do much differently but would be wary of office politics. My experience might be unique because I was working on a project with another company to deliver a project for a client. I was oblivious to the office politics at play, and as a result, someone I thought was a friend in the team convinced the client to drop our company. Furthermore, I was too junior at the time, and it was argued that I was taking up too much of this person's time. On the other hand, I was told to ask many questions as this person had a lot more knowledge than me, and as a result, I learnt a tremendous amount. I didn't find out about this until months after we lost the client. It was quite devastating to hear that as a result of doing my job and trying to learn as much as I could, I cost the company a client and a project I enjoyed working on. This was a tough lesson to learn: not everyone is your friend, especially when their motives are not aligned with yours (this can happen when they are from another company competing for work). If I were to do anything differently, I would be more weary of other people's motives, especially when they are not in the same company, and even still be aware that not everyone out there will be nice behind your back. It is a lesson you must learn, and you will, at some point, learn it. As a result, I have been able to spot people that might do this and not take it to heart as much. If you take these things to heart, it will never end well; more than likely, it's never personal.

What You Should Know About Your First Year

Your first year is tough if you haven't already gathered from this book. You will stumble, struggle, and wonder why you selected this field. When those lows hit, come back to this book, consider why you got into the field and be grateful for the experiences. They won't last forever, and you will get through them. One of the experiences I keep having, time and time again, is when I take that next step in my career, it's extraordinarily hard. When I became a tech lead for the first time, I was freaking out; I constantly had no time to do anything. I was working extra hours to make up for everything I couldn't get to. After my first two weeks, I told my father-in-law, 'I severely underestimated how hard a tech lead role is.' He said, 'Come back to me in a month.' After a month, I had it down pat and was on top of almost everything. There were still challenges, but I was no longer drowning in the ocean but happily floating. You will get there; it just takes time. You will feel that it's too much at the time, but you will get there. Furthermore, you will question everything, especially your intelligence, and think you shouldn't be in this field, but you will get there. It all takes time! Before I started university, I still remember that I couldn't fathom how Facebook was built even before I knew what a variable was. Slowly, over time, I put the pieces together and gradually gained knowledge. It wasn't until years later that I thought, 'Wow, I know how Facebook is built now, maybe not exactly, but I can piece it all together.' I thought back to how I didn't know anything, and now I know a lot more. I went from knowing nothing to being highly successful in the field. It just takes time. If you get nothing from this book, remember this one point: it just takes time!

The Key Takeaways You Can Use to Have a Successful Career

If you have skipped to this section of the book, good on you; here is a summary of the whole thing.

Ask Plenty of Questions

Asking questions is the most important thing you can do as a new developer. The more you ask, the less time you waste trying to figure things out. You should still try to figure things out after asking all your questions because that is how you learn the most effectively. If something isn't making sense, you aren't dumb; you just need to ask a few more questions.

Work Hard, but Work Smart

Don't work more than eight hours for your job. If you want to do extra, learn something new or build something on the side.

Take Care of Your Health

Exercise, eat well, reduce stress and sleep well. These are essential to staying happy, healthy and productive for years to come. Do not skip this.

Software Development Isn't about the Technology

This field is complex from a social and difficulty standpoint. You will have many challenging people in your teams and workplace, and as a result, the code will be the easier beast. You must learn to work with people, respect others, empathise and give people a break. Furthermore, you pick up the skills to complete your tickets, but the skill of working with others is much more important. The team is only as strong as its weakest link, and if that weakest link is someone who doesn't want to work with others, the entire team will suffer.

Software Development Is Hard

Many people outside the field believe software is extremely hard, and it is hard. Bugs can be next to impossible to find and fix; no matter what you try, this bit of code won't work, and someone else breaks all the features with a small code change. So many factors make software development hard, and you must accept it, slow down, breathe and push on. You will find a solution eventually; it just takes time. If you are pressured by deadlines, it's up to you to communicate why it's taking so long and if there is anything that can be done to reduce this time. Software development is hard, so working well with your team is important to get things done.

Understand Your Value

You are valuable, have unique skills, work hard, and continue to learn and improve. Do not let companies take advantage of you, push you around, pay you less than your worth or force you to work overtime. Stand up respectfully for yourself and your team.

Respect, Empathy, Reframing

Respect everyone around you. You may not like them, but they have got to where they are for a reason.

Empathise with people. Most people don't intentionally harm others. People go through rough times and take it out on others. If someone is rude or angry, consider they might be dealing with tough personal circumstances; it's no excuse, but it can be a reason for their behaviour. Empathise with them, reach out and try to help however you can.

Reframe your mindset. Reframe your negative thoughts about the project, people, or company into positive thoughts to get the most out of your situations. There is always a bright side.

Be Loyal to Yourself

Loyalty is something that I struggle with because I am too loyal. I want to avoid rocking the boat or making a fuss with anyone because I wish to be liked. Most people are the same; they will sit in silence and often be taken advantage of over speaking up. The only people you should be loyal to are yourself and your loved ones. You should never be loyal to a job because they won't be loyal to you when you most need it. If you let loyalty control whether you stay in a poor situation, take less money, or skip a promotion, you will lose out eventually. You will resent yourself and your job and take this out on others. Loyalty is one of the hardest things to overcome, but in today's world, companies will never choose you over their bottom line. Take it from me; I was made redundant one week after moving into a new house.

Companies will use the lure of promotion to tempt you to stay, and they will feed you lies that if you work hard and put in the extra hours, you will go far, get promoted and make lots of money. Then, when it comes time to give you these benefits, they make up excuses.

These are the classics:

- 'You haven't been here long enough.'
- 'You would be the highest-paid person in the company.'
- 'We can't afford it.'

All these lines are used to scapegoat you into thinking you have done a good job and that they simply can't give you the raise and promotion. These are all lies; if the business can't afford to pay you, it's not well-run. Loyalty will keep you at these companies, and these companies will use this against you. They drive you to work longer, harder, and for less money, all for the team's greater good. They even do it in more subtle ways by playing you off against other people in the same team who are more loyal. This is your career; you make the decisions and are the most important person in it. Make whatever decision is right for you, take any path you want, and never settle for less.

BOOK RECOMMENDATIONS

Cal Newport
- *Deep Work*
- *So Good They Can't Ignore You*

Robert C. Martin
- *Clean Code*
- *Clean Architecture*

Andy Hunt and Dave Thomas
- *The Pragmatic Programmer*

Timothy Lister and Tom DeMarco
- *Peopleware: Productive Projects and Teams*

Camille Fournier
- *The Manager's Path*

Eliyahu M. Goldratt
- *The Goal*

Steve McConnell
- *Code Complete*